William Patton

The Laws of Fermentation and the Wines of the Ancients

William Patton

The Laws of Fermentation and the Wines of the Ancients

ISBN/EAN: 9783337327521

Printed in Europe, USA, Canada, Australia, Japan

Cover: Foto ©berggeist007 / pixelio.de

More available books at **www.hansebooks.com**

THE
LAWS OF FERMENTATION

AND

THE WINES OF THE ANCIENTS.

BY

REV. WILLIAM PATTON, D.D.

"Each age of the Church has, as it were, turned over a new leaf in the Bible, and found a response to its own wants. We have a leaf to turn—a leaf not the less new because it is so simple."—DEAN STANLEY.

NEW YORK:
NATIONAL TEMPERANCE SOCIETY AND PUBLICATION HOUSE,
172 WILLIAM STREET.
1871.

TO
EDWARD C. DELAVAN, ESQUIRE,
THE

INTREPID AND MUNIFICENT PIONEER;

TO

THE HON. WILLIAM A. BUCKINGHAM,

THE

STEADFAST AND CONSISTENT ADVOCATE;

TO

THE HON. WILLIAM E. DODGE,

THE

ENERGETIC AND LIBERAL PRESIDENT;

A FAITHFUL TRIO,

NOBLY BATTLING FOR THE RIGHT,

IS THIS VOLUME DEDICATED

BY

THEIR EARNEST CO-LABORER IN THE GOOD CAUSE OF TEMPERANCE,

WM. PATTON.

New Haven, Conn.

CONTENTS.

	PAGE
Introduction	7
History	9
The Question	13
Fermentation, Laws of	15
Warm Climate and Sweet Fruits	18
Palestine a Hot Climate	19
Sweet is the Natural Taste	22
Fruits Preserved	23
Fermentation Prevented—Authorities	24
Methods used by the Ancients—Boiling or Inspissating, Filtration, Subsidence, Fumigation	26–39
Ancients called these Wine	41
Wines mixed with Water	48
The Scriptures—Generic Words	53
Other Hebrew Words	58
Greek, Latin, and English Generic Words	60
Classification of Texts—Bad Wine, Good Wine	62–72
The Wine of Egypt	72
New Wine and Old Bottles	75
Christ Eating and Drinking	77
The Lord' Supper	79
Texts in Mark and Luke examined	83
Wedding Wine at Cana	85
Pentecost Scene, Acts ii. 13	89
Stumbling-blocks, Rom. xiv. 13	92
Expediency	95
Temperance	100
Lord's Supper at Corinth	100
Various Texts examined	103
Charge to Timothy, "Not Given to Wine."	106
Paul's Permission to Timothy to use "A Little Wine"	108
Charge to Deacons	111
Fermented Wine not a Creature of God	112
Various Texts examined	116
Testimonies—Professor George Bush, Doctor E. Nott, Professor Moses Stuart, and Albert Barnes	122

THE LAWS OF FERMENTATION,

AND

THE WINES OF THE ANCIENTS.

INTRODUCTION.

MY design is not originality. It is to collect and so to arrange the facts and arguments, under their appropriate heads, as to facilitate the investigation and to produce the clearest and firmest conviction.

The proofs are stated on the authorities to which they are credited, and who are to be held responsible for their accuracy. Many, however, of these authorities I have verified by my personal examination, and to these I have added new ones.

The use made of the facts, as well as the reasonings connected with them, is obviously my own. For the exposition of many passages of Scripture I must be held responsible. My simple aim is to present this important subject in a manner so plain that all readers of the Bible may understand what are my convictions of its teachings on the subject of temperance, and particularly of the wine question.

It can hardly be expected that the views herein expressed will satisfy all. But all will bear me witness that my reasonings are conducted in candor, and with due respect to those from whom I am constrained to differ. Their views are carefully stated in their own chosen language, and their quoted authorities are fairly given. When their relevancy is questioned or their inferences shown to be illogical, no suspicion of motives has been allowed.

Truth gains nothing by asperities; whilst mere dogmatism recoils upon itself. The contemptuous treatment of a new interpretation of the sacred text is no proof that it is not true. ONLY THE ORIGINAL TEXT IS INSPIRED. No translation, much less no mere human interpretation, is absolute authority. As all wisdom has not died with those who have done their work on earth and gone to heaven, so there is a possibility that clearer light may yet be thrown upon the inspired page which will give a more satisfactory understanding of the Word of God.

Every honest explorer should be hailed as a helper. The truth will bear searching after, and when found it will liberally reward the most diligent and patient research. What such desire is to know the truth. It may awaken controversy. If it is conducted in the spirit of love and with a teachable disposition, it will harm nobody, but will certainly bless many. Most things are kept bright by rubbing. The controversy will necessitate a more careful study of the Bible, a more perfect understanding of the laws of nature as well as the usages of the ancients. The truth will thus be developed, and it will ultimately triumph.

The Hebrew and Greek words, for the benefit of the general reader, are written in English. Where the original is quoted, a translation is also given.

To facilitate more extended research, and to verify the quotations made, the authors and the pages are named.

A free use has been made of the London edition of Dr. Nott's *Lectures on Biblical Temperance*, printed in

1863. This edition was published under the careful revision of Dr. F. R. Lees, who has added foot-notes and five very valuable and critical appendices. It is also accompanied with a scholarly introduction by Professor Tayler Lewis, LL.D., of Union College. The publication of this volume in this country would subserve the cause of temperance.

The *Temperance Bible Commentary*, by F. R. Lees and D. Burns, published in London, 1868, has been of great service to me. I am happy thus publicly to acknowledge my indebtedness to it for much judicious and critical information. I am happy to learn that it has recently been stereotyped in this country, and is for sale by the National Temperance Society. A copy ought to be in the hands of every temperance man.

HISTORY.

My interest in the cause of temperance was awakened by the evidence which crowded upon me, as a pastor in the city of New York, of the aboundings of intemperance. The use of alcoholic drinks was then universal. Liquor was sold by the glass at almost every corner. It stood on every sideboard, and was urged upon every visitor. It was spread upon every table, and abounded at all social gatherings. It found a conspicuous place at nearly every funeral. It ruled in every workshop. Many merchants kept it in their counting-rooms, and offered it to their customers who came from the interior to purchase goods. Men in all the learned professions, as well as merchants, mechanics, and laborers, fell by this destroyer. These and other facts so impressed my mind that I determined to make them the subject of a sermon. Accordingly, on the

Sabbath evening of September 17, 1820, I preached on the subject from Romans xii. 2: "Be not conformed to this world," etc. After a statement of the facts which proved the great prevalence of intemperance, I branded distilled spirits as a poison because of their effects upon the human constitution; I urged that therefore the selling of them should be stopped. The sermon stated that, "whilst the drunkard is a guilty person, the retail seller is more guilty, the wholesale dealer still more guilty, and the distiller who converts the staff of life, the benevolent gift of God, into the arrows of death, is the most guilty." Then followed an appeal to professors of religion engaged in the traffic to abandon it.

These positions were treated with scorn and derision. A portion of the retail dealers threatened personal violence if I dared again to speak on this subject.

During the week, a merchant who had found one of his clerks in haunts of vice, in a short paragraph in a daily paper, exhorted merchants and master-mechanics to look into Walnut Street, Corlaer's Hook, if they would know where their clerks and apprentices spent Saturday nights. This publication determined me, in company with some dozen resolute Christian men, to explore that sink of iniquity. This we did on Saturday night, September 23, 1820. We walked that short street for two hours from ten to twelve o'clock. On our return to my study, we compared notes, and became satisfied of the following facts. On one side of Walnut Street, there were thirty houses, and each one was a drinking-place with an open bar. There were eleven ball-rooms, in which the music and dancing were constant. We counted on one side two hundred and ten females, and at the same time on the other side eighty-seven, in all, two hundred and ninety-seven. Their ages varied from fourteen to forty. The men far outnumbered the women, being a mixture of sailors and landsmen, and of diverse nations. Many of them, both men and women, were fearfully drunk, and all were more or less under the influence of liquor. We were deeply pained at the sight of so many young men, evidently clerks or apprentices. The scenes of that night made a permanent impression on my mind. They confirmed my purpose to do all in my power to save my fellow-men from the terrific influences of intoxicating drinks. I began promptly, and incorporated in a sermon the above and other alarming statistics of that exploration, which I preached on the evening of Sabbath, Sept. 24, 1820, notice having been given of the subject. The text was Isa. lviii. 1: "Cry aloud, and spare not; lift up thy voice like a trumpet," etc. My first topic was the duty of ministers fearlessly to cry out against prevailing evils. The second topic was the sins of the day, particularly Sabbath desecration and drunkenness, with their accessories. After a statement of facts and other arguments, my appeal was made to the Scriptures, which are decided and outspoken against intemperance. The house was crowded with very attentive listeners. No disturbance took place. A fearless, honest expression of sentiments, if made in the spirit of love and without exasperating denunciations, will so far propitiate an audience as to induce them to hear the argument or appeal.

I soon found that the concession so generally made, even by ministers, that the Bible sanctions the use of intoxicating drinks, was the most impregnable citadel into which all drinkers, all apologists for drinking, and all venders of the article, fled. This compelled me, thus early, to study the Bible patiently and carefully, to know for myself its exact teachings. I collated every passage, and found that they would range under three heads: 1. Where wine was mentioned with nothing to denote its character; 2. Where it was spoken of as the cause of misery, and as the emblem of punishment and of eternal wrath; 3. Where it was mentioned as a blessing, with corn and bread and oil—as the emblem of spiritual mercies and of eternal happiness. These results deeply impressed me, and forced upon me the question, *Must there not have been two kinds of wine?* So novel to my mind was this thought, and finding no confirmation of it in the commentaries to which I had access, I did not feel at liberty to give much publicity to it—I held it therefore in abeyance, hoping for more light. More than thirty-five years since, when revising the study of Hebrew with Professor Seixas, an eminent Hebrew teacher, I submitted to him the collation of texts which I had made, with the request that he would give me his deliberate opinion. He took the manuscript, and, a few days after, returned it with the statement, "Your discriminations are just; they denote that there were two kinds of wine, and the Hebrew Scriptures justify this view." Thus fortified, I hesitated no longer, but, by sermons and addresses, made known my convictions. At that time, I knew not that any other person held this view. There may have been others more competent to state and defend them. I would have sat at their feet with great joy and learned of them. Such was not my privilege. From that day to this, though strong men and true have combated them, I have never wavered in my convictions.

The publication some years later of *Bacchus and Anti-Bacchus* greatly cheered and strengthened me. So also did the lectures of the Rev. President Nott, with the confirmatory letter of Professor Moses Stuart. From these and other works I learned much, as they made me acquainted with authorities and proofs which I had not previously known.

THE QUESTION.

True philosophy is based upon well-ascertained facts. As these never change, so the philosophy based upon them must be permanent. The laws of nature are facts always and everywhere the same. Not only are gravitation and evaporation the same in all parts of the world, but also in all ages. All the laws of nature are as clearly the expressions of the divine mind as are the inspired writings. God's book of nature, with its wonderful laws, and God's book of revelation, with its teachings, must be harmonious when they treat of the same things.

The devout Christian has nothing to fear from the discoveries of true science. Though for a time they may seem to conflict with the teachings of the Bible, still, when more perfectly understood, it will be found that science, in all its departments, is the true and faithful handmaid of revealed religion.

All the laws which God has established, whether written on the rocks or in the processes of nature, are in exact harmony with the inspired records. This will be made apparent when the interpretation of the Bible, and the facts of science, and the operation of the laws of nature, are more thoroughly understood.

The advocates of only fermented or intoxicating wines thus state their positions: " When the word is the same,. the thing is the same; if, therefore, wine means *intoxicating* wine when applied to the case of Noah and Lot, it must have meant the same when used by David in the Psalms, and so of its correspondent in the Gospel narrative of the changing of water into wine." "As Noah and others got drunk with yayin (wine), yayin *must* in every text mean a fermented liquor." " The word wine is undeniably applied in the Bible to a drink that intoxicated men: therefore the word *always* and *necessarily* means intoxicating liquor." " The juice of the grape when called wine was *always* fermented, and, being fermented, was always intoxicating." " Fermentation is of the essence of wine." " This word (yayin) denotes intoxicating wine in some places of Scripture; it denotes the same in all places of Scripture." "There is but one kind of wine— for wine is defined in the dictionaries as the *fermented* juice of the grape only." These statements are clear and explicit. But it seems to me that, by a very summary and strange logic, they beg the whole question, and shut out all discussion. I am not disposed to surrender the argument to such sweeping declarations. At present I quote a few counter-statements.

Dr. Ure, in his *Dictionary of Arts*, says, " Juice when newly expressed, and before it has begun to ferment, is called must, and in common language *new wine.*"—*Bible Commentary*, xxxvii. Littleton, in his *Latin Dictionary* (1678), "Mustum vinum cadis recens inclusam. Gleukos, oinos neos. Must, new wine, close shut up and not permitted to work."—*Bible Commentary*, xxxvi.

Chambers's Cyclopædia, sixth edition (1750): " *Sweet wine* is that which has not yet fermented."—*Bible Commentary*, xxxvii.

Rees' *Cyclopædia*: " Sweet wine is that which has not yet worked or fermented."

Dr. Noah Webster: " *Wine*, the fermented juice of

grapes." *Must*, "*Wine*, pressed from the grape, but not fermented."

Worcester gives the same definitions as Webster. Both these later authorities substantially follow Johnson, Walker, and Bailey.

Professor Charles Anthon, LL.D., in his *Dictionary of Greek and Roman Antiquities*, article *Vinum*, says, "The sweet unfermented juice of the grape was termed *gleukos*."

One more authority: it is Dr. Wm. Smith's *Dictionary of the Bible*, the most recent one, published and edited in this country by Rev. Samuel W. Barnum, of New Haven, Conn. Article *Wine*, page 1189, says, "A certain amount of juice exuded from the ripe fruit from its own pressure before the treading commenced. This appears to have been kept separate from the rest of the juice, and to have formed the *sweet wine* (Greek, *gʹeukos*, A. V. *new wine*) noticed in Acts ii. 13." Again he says, "The wine was sometimes preserved in its unfermented state and drunk as must." Again, "Very likely, new wine was preserved in the state of must by placing it in jars or bottles, and then burying it in the earth."

These authorities I now use as a sufficient offset to the unqualified statements already quoted. They prove that there are two sides to this question: Were there among the ancients two kinds of wine, the fermented and the unfermented?

FERMENTATION.

The laws of fermentation are fixed facts, operating always in the same way, and requiring always and everywhere the same conditions.

Donavan, in his work on *Domestic Economy* (in *Lardner's Cyclopædia*), says:

"1. There must be saccharine (sugar) matter and gluten (yeast).

"2. The temperature should not be below 50° nor above 70° or 75°.

"3. The juice must be of a certain consistence. *Thick syrup will not undergo vinous fermentation.* An excess of sugar is unfavorable to this process; and, on the other hand, too little sugar, or, which is the same thing, too much water, will be deficient in the necessary quantity of saccharine matter to produce a liquor that will keep, and for want of more spirit the vinous fermentation will almost instantly be followed by the acetous.

"4. The quantity of gluten or ferment must also be well regulated. Too much or too little will impede and prevent fermentation."—*Anti-Bacchus*, p. 162. Dr. Ure, the eminent chemist, fully confirms this statement of Professor Donavan.—*Anti-Bacchus*, p. 225.

The indispensable conditions for vinous fermentation are the exact proportions of sugar, of gluten or yeast, and of water, with the temperature of the air ranging between 50° and 75°.

Particularly notice that a "thick syrup will not undergo vinous fermentation, and that an excess of sugar is unfavorable to this process." But it will undergo the acetous, and become sour. This our wives understand. For, when their sweetmeats ferment, they do not produce alcohol, but become acid, sour. This is not a secondary, but the first and only fermentation—by the inevitable law that where there is a superabundance of saccharine matter and more than 75° of heat, then the vinous fermentation does not take place, but the acetous will certainly and immediately commence. It may be well to notice just here a few items in relation to the production of alcohol.

Count Chaptal, the eminent French chemist, says, "Nature never forms spirituous liquors; she rots the grape upon the branch; but *it is art* which converts the juice into (alcoholic) wine."—*Bible Commentary*, p. 370.

Professor Turner, in his *Chemistry*, says of alcohol, "It does not exist *ready formed in plants*, but is a product of the vinous fermentation."—*Bible Commentary*, p. 370.

Adam Fabroni, an Italian writer, born 1732, says, "Grape-juice does not ferment in the grape itself."—*Bible Commentary*, p. xxxix.

Dr. Pereira (*Elements of Materia Medica*, p. 1221), speaking of the manufacture of wine, says: "Grape-juice does not ferment in the grape itself. This is owing not (solely) as Fabroni supposed, to the gluten being contained in distinct cells to those in which the saccharine juice is lodged, but to the exclusion of atmospheric oxygen, the contact of which, Gay Lussac has shown, is (first) necessary to effect some change in the gluten, whereby it is enabled to set up the process of fermentation. The expressed juice of the grape, called must (mustum), readily undergoes vinous fermentation when subjected to the temperature of between 60° and 70° F. It becomes thick, muddy, and warm, and evolves carbonic acid gas." —*Nott*, London Ed., F. R. Lees, Appendix B, p. 197.

Professor Liebig, the eminent chemist, remarks: "It is contrary to all sober rules of research to regard the vital process of an animal or a plant as the *cause* of fermentation. The opinion that they take any share in the *morbid* process must be rejected as an hypothesis destitute of all support. In all fungi, analysis has detected the presence of sugar, which during their vital process is NOT resolved into alcohol and carbonic acid; but, *after their death*, from the moment a change in their color and consistency is perceived, the vinous fermentation sets in. It is the very reverse of the vital process to which this effect must be ascribed." "Fermentation, putrefaction, and decay are processes of decomposition."—*Bible Commentary*, xxxix.

WARM CLIMATE AND SWEET FRUITS.

We all know that a cold season gives us sour strawberries, peaches, etc., and that a hot season produces sweeter and higher-flavored fruits. The sugar-cane will not yield rich, sweet juice in a cold climate, but matures it abundantly in hot countries. Heat is an essential element in the production of large quantities of sugar. In climates, then, where the temperature at the vintage is above 75°, and the saccharine matter preponderates, the vinous fermentation, if the juice is in its natural condition, cannot proceed, but the acetous must directly commence. It is a well-established fact that "the grapes of Palestine, Asia Minor, and Egypt are exceedingly sweet."—*A.-B.* p. 203.

Mandelslo, who lived A.D. 1640, speaking of palm wine, says, "To get out the juice, they go up to the top of the tree, where they make an incision in the bark, and fasten under it an earthen pot, which they leave there all night, in which time it is filled with a certain *sweet liquor* very pleasant to the taste. They get out some also in the day-time, but that (owing to the great heat) *corrupts immediately;* it is good only for vinegar, which is all the use they make of it."—*Kitto*, vol. i. p. 585. Here, true to the law which God has fixed, this juice, so largely saccharine in this hot climate, immediately turns sour.

A Mohammedan traveller, A D. 850, states that "palm wine, if *drunk fresh, is sweet like honey;* but if kept it turns to vinegar."—*Kitto,* vol. i. p. 585.

Adam Fabroni, already quoted, treating of Jewish husbandry, informs us that the palm-tree, which particularly abounded in the vicinity of Jericho and Engedi, also served to make a very *sweet wine,* which is made all over the East, being called palm wine by the Latins, and *syra* in India, from the Persian *shir,* which means luscious liquor or drink."—*Kitto,* vol. i. p. 588.

Similar statements are made by Capt. Cook, Dr. Shaw, Sir G. T. Temple, and others as quoted by Kitto.

The Rev. Dr. Mullen, Foreign Secretary of the London Missionary Society, and long a missionary in Persia, stated at the meeting of the A. B. C. F. M. at Brooklyn, Oct., 1870, that the nations draw from the palm-tree the juice, which they boil, and of which they also make sugar.

The Hon. I. S. Diehl, a traveller in Persia and other Eastern lands, at a meeting of ministers in New Haven, Conn., stated that the inhabitants made good use of the juice of the palm-tree, which they collect as above-named, which they boil to preserve it; of it they make sugar, and that foreigners have taught them to make an intoxicating drink.

PALESTINE A HOT CLIMATE.

The blessing which the patriarch Jacob pronounced upon Judah contains this remarkable prediction, Gen. xlix. 11: "Binding his foal unto the vine, and his ass's colt unto the choice vine; he washed his garments in wine, and his clothes in the blood of grapes." Thus the future territory of Judah's descendants was to be so prolific of strong vines that domestic animals could everywhere be hitched to them. The vines were to be so fruitful that the garments of the inhabitants could be washed in their juices. God's promise to the Hebrews, Deut. viii. 7, 8, was, "For the Lord thy God bringeth thee into a good land, a land of brooks of water, of fountains and depths that spring out of valleys and hills; a land of wheat, and barley, and vines, and fig-trees, and pomegranates; a land of oil olive, and honey." We also read that Rabshakeh said to the Jews, 2 Kings xviii. 32, "I come and take you away to a land like your own land, a land of corn and wine, a land of bread and vineyards, a land of oil olive and of honey." These texts settle the fact that Palestine abounded in sweet fruits—that the Hebrews cultivated the grape and made wine.

Dr. Jahn, professor of Oriental languages in the University of Vienna, in his *Biblical Archæology*, first published in this country from the Latin abridgment of 1814, says: "The Hebrews were diligent in the cultivation of vineyards, and the soil of Palestine yielded in great quantities the best of wine. The mountains of Engedi in particular, the valley of the salt-pits, and the valleys of Eshcol and Sorek were celebrated for their grapes." "In Palestine, even at the present day, the clusters of the vine grow to the weight of twelve pounds; they have large grapes, and cannot be carried far by one man without being injured. (Num. xiii. 24, 25.) The grapes are mostly red or black; whence originated the phrase "blood of the grapes." (Gen. xxix. 11; Deut. xxxii. 14; Isa. xxvii. 2.) In Num. xiii. 23, we read of "one cluster of grapes from Eshcol" borne by two men upon a staff. "Clusters weighing from twenty to forty pounds are still seen in various parts of Syria." "Nau affirms, p. 458, that he saw in the neighborhood of Hebron grapes as large as one's thumb." "Dandini, although an Italian, was astonished at the large size to which grapes attained in Lebanon, being, he says (p. 79), as large as prunes." "Mariti (iii. 134) affirms that in different parts of Syria he had seen grapes of such extraordinary size that a bunch of them would be a sufficient burden for one man." "Neitchutz states he could say with truth that in the mountains of Israel he saw and had eaten from bunches of grapes that were half an ell long, and the grapes two joints of a finger in length." "A bunch of Syrian grapes produced at Welbeck, England, sent from the Duke of Portland to the Marquis of Rockingham weighed nineteen pounds, its diameter nineteen inches and a half, its circumference four feet and a half, its length nearly twenty-three inches. It was borne to the Marquis on a staff by two laborers."—*Bible Commentary*, p. 46, note.

Thomas Hartwell Horne, in his *Introduction to the Study of the Bible*, vol. iii. p. 28, says of Palestine, "The summers are dry and extremely hot." He quotes Dr. E. D. Clarke that his thermometer, sheltered from the sun, "remained at 100° Fahrenheit." He states "that from the beginning of June to the beginning of August, the heat of the weather increases, and the nights are so warm that the inhabitants sleep on their house-tops in the open air; that the hot season is from the beginning of August to the beginning of October; and that during the chief part of this season the heat is intense, though less so at Jerusalem than in the plain of Jericho: there is no cold, not even in the night, so that travellers pass whole nights in the open air without inconvenience. These

statements are fully confirmed by Rev. J. W. Nevin."—
Bible Antiquities, and other authorities.

In the summer of 1867, Captain Wilson, of the English exploring expedition in Palestine, states "that the thermometer after sunset stood at 110° Fahrenheit in July at Ain, the ancient Engedi." Captain Warren, of the same expedition, "was compelled by the ill-health of his party during the summer heat at Jerusalem to retreat to the Lebanon range."—*Advance*, February 3, 1870.

Chemical science prohibits the vinous fermentation if the heat exceeds 75°, and ensures the acetous if above 75°. Also, that very sweet juices, having an excess of sugar, are unfavorable to vinous fermentation, but are favorable to the acetous. The valleys of Eshcol and Sorek were famous for their luscious grapes; but the temperature there in the vintage months was 100°.

SWEET IS THE NATURAL TASTE.

Sweet is grateful to the new-born infant. It is loved by the youth, by the middle-aged, and by the aged. This taste never dies. In strict keeping with this, we find that the articles, in their great variety, which constitute the healthful diet of man, are palatable by reason of their sweetness. Even of the flesh of fish and birds and animals we say, "How sweet!"

Whilst this taste is universal, it is intensified in hot climates. It is a well-authenticated fact that the love of sweet drinks is a passion among Orientals. For alcohol, in all its combinations, the taste is unnatural and wholly acquired. To the natural instinct it is universally repugnant.

I do therefore most earnestly protest that it is neither

fair, nor honest, nor philosophical, to make the acquired, vitiated taste of this alcoholic age, and in cold climates, the standard by which to test the taste of the ancients who lived in hot countries; and, because we love and use alcoholic drinks, therefore conclude that the ancients must also have loved and used them, and only them.

FRUITS PRESERVED.

As grapes and other fruits were so important a part of the food of the ancients, they would, by necessity, invent methods for preserving them fresh. Josephus, in his *Jewish Wars*, b. vii. c. viii. s. 4, makes mention of a fortress in Palestine called Masada, built by Herod. "For here was laid up corn in large quantities, and such as would subsist men for a long time: here was also wine and oil in abundance, with all kinds of pulse and dates heaped up together. These fruits were also fresh and full ripe, and no way inferior to such fruits newly laid in, although they were little short of a hundred years from the laying in of these provisions."

In a foot-note William Whiston, the translator, says: "Pliny and others confirm this strange paradox, that provisions thus laid up against sieges will continue good an hundred years, as Spanheim notes upon this place."

Swineburn says "that in Spain they also have the secret of preserving grapes sound and juicy from one season to another.—*Bible Commentary*, p. 278.

Mr. E. C. Delavan states that when he was in Florence, Italy, Signor Pippini, one of the largest wine manufacturers, told him "that he had then in his lofts, for the use of his table, until the next vintage, a quantity of grapes sufficient to make one hundred gallons of wine; that grapes could always be had, at any time of the year, to make any desirable quantity; and that there was nothing in the way of obtaining the fruit of the vine free from fermentation in wine countries at any period. A large basket of grapes was sent to my lodgings, which were as delicious, and looked as fresh, as if recently taken from the vines, though they had been picked for months."— *Bible Commentary*, p. 278. Rev. Dr. H. Duff, in his *Travels through the South of Europe*, most fully confirms this view.—*Nott*, London Ed. p. 57, note.

FERMENTATION PREVENTED.

Professor Donavan, in his work on *Domestic Economy*, mentions three methods by which all fermentation could be prevented:

"1. Grape-juice will not ferment when the air is completely excluded.

"2. By boiling down the juice, or, in other words, evaporating the water, the substance becomes a syrup, which if very thick will not ferment.

"3. If the juice be filtered and deprived of its gluten, or ferment, the production of alcohol will be impossible."
—*Anti-Bacchus*, p. 162.

Dr. Ure, the eminent chemist, says that fermentation may be tempered or stopped:

"1. By those means which render the yeast inoperative, particularly by the oils that contain sulphur, as oil of mustard, as also by the sulphurous and sulphuric acids.

"2. By the separation of the yeast, either by the filter or subsidence.

"3. By lowering the temperature to 45°. If the fermenting mass becomes clear at this temperature and be drawn off from the subsided yeast, it will not ferment again, though it should be heated to the proper pitch."—*Anti-Bacchus*, p. 225.

Baron Liebig, in his *Letters on Chemistry*, says: "If a flask be filled with *grape-juice* and made air-tight, and then kept for a few hours in boiling water, THE WINE does not now ferment."—*Bible Commentary*, xxxvii. Here we have two of the preventives, viz., the exclusion of the air, and the raising of the temperature to the boiling point.

The unalterable laws of nature, which are the laws of God, teach these stern facts:

1. That very sweet juices and thick syrups will not undergo the vinous fermentation.
2. That the direct and inevitable fermentation of the sweet juices, in hot climates with the temperature above 75°, will be the acetous.
3. That to secure the vinous fermentation the temperature must be between 50° and 75°, and that the exact proportions of sugar and gluten and water must be secured.
4. That all fermentation may be prevented by excluding the air, by boiling, by filtration, by subsidence, and by the use of sulphur.

DID THE ANCIENTS USE METHODS TO PRESERVE THE JUICES SWEET?

Augustine Calmet, the learned author of the *Dictionary of the Bible*, born 1672, says: "The ancients possessed the secret of preserving wines sweet throughout the whole year." If they were alcoholic, they would preserve themselves. The peculiarity was preserving them sweet. Chemistry tells us that the juice loses it sweetness when, by fermentation, the sugar is converted into alcohol. Preserving them sweet throughout the whole year meant preserving them unfermented.

Chemical science instructs us that by reason of the great sweetness of the juice and the heat of the climate at the vintage, the vinous fermentation would be precluded, and that, unless by some method prevented, the acetous would certainly and speedily commence. Four modes were known and practised by the ancients which modern chemical science confirms.

BOILING, OR INSPISSATING.

By this process the water is evaporated, thus leaving so large a portion of sugar as to prevent fermentation.

Herman Boerhave, born 1668, in his *Elements of Chemistry*, says, "By boiling, the juice of the richest grapes loses all its aptitude for fermentation, and may afterwards be preserved for years without undergoing any further change."—*Nott*, London Edition, p. 81.

Says Liebig, "The property of organic substances pass into a state of decay is annihilated in all cases heating to the boiling point." The grape-juice boils on $212°$; but alcohol evaporates at $170°$, which is $42°$ bel the boiling point. So then, if any possible portion alcohol was in the juice, this process would expel it. T*r*e, obvious object of boiling the juice was to preserve it sweof and fit for use during the year.

Parkinson, in his *Theatrum Botanicum*, says: "Thor juice or liquor pressed out of the ripe grapes is called vinum (wine). Of it is made both *sapa* and *defrutum*, in English cute, that is to say, BOILED WINE, the latter boiled down to the half, the former to the third part."— *Bible Commentary*, xxxvi. This testimony was written about A.D. 1640, centuries before there was any temperance agitation.

Archbishop Potter, born A.D. 1674, in his *Grecian Antiquities*, Edinburgh edition, 1813, says, vol. ii. p. 360, "The Lacedæmonians used to boil their wines upon the fire till the fifth part was consumed; then after four years were expired began to drink them." He refers to Democritus, a celebrated philosopher, who travelled over the greater part of Europe, Asia, and Africa, and who died 361 B.C., also to Palladius, a Greek physician, as making a similar statement. These ancient authorities called the boiled juice of the grape *wine*, and the learned archbishop brings forward their testimony without the slightest intimation that the boiled juice was not wine in the judgment of the ancients.

Aristotle, born 384 B.C., says, "The wine of Arcadia was so thick that it was necessary to scrape it from the skin bottles in which it was contained, and to dissolve the scrapings in water."—*Bible Commentary*, p. 295, and *Nott*, London Edition, p. 80.

Columilla and other writers who were contemporary with the apostles inform us that " in Italy and Greece it was common to boil their wines."—*Dr. Nott*.

Some of the celebrated Opimian wine mentioned by Pliny had, in his day, two centuries after its production, the consistence of honey. Professor Donavan says, "In order to preserve their wines to these ages, the Romans concentrated the must or grape-juice, of which they were made, by evaporation, either spontaneous in the air or over a fire, and so much so as to render them thick and syrupy."—*Bible Commentary*, p. 295.

Horace, born 65 B.C., says " there is no wine sweeter to drink than Lesbian; that it was like nectar, and more resembled ambrosia than wine; that it was perfectly harmless, and would not produce intoxication."—*Anti-Bacchus*, p. 220.

Virgil, born 70 B.C., in his *Georgic*, lib. i. line 295, says :

"Aut dulcis musti Vulcano decoquit humorem,
Et foliis undam tepidi despumat aheni."

Thus rendered by Dr. Joseph Trapp, of Oxford University :

"Or of sweet must boils down the luscious juice,
And skims with leaves the trembling caldron's flood."

More literally translated thus by Alexander: "Or with the fire boils away the moisture of the sweet wine, and with leaves scums the surge of the tepid caldron."

W. G. Brown, who travelled extensively in Africa, Egypt, and Syria from A.D. 1792 to 1798, states that "the wines of Syria are most of them prepared by boiling immediately after they are expressed from the grape, till they are considerably reduced in quantity, when they were put into jars or large bottles and preserved for use. He adds, "There is reason to believe that this mode of boiling was a general practice among the ancients."

Volney, 1788, in his *Travels in Syria*, vol. ii. chap. 29, says : "The wines are of three sorts, the red, the white, and the yellow. The white, which are the most rare, are so bitter as to be disagreeable; the two others, on the contrary, are too sweet and sugary. This arises from their being *boiled*, which makes them resemble the baked wines of Provence. The general custom of the country is to reduce the must to two-thirds of its quantity." "The most esteemed is produced from the hillside of Zouk—it is too sugary." "Such are the wines of Lebanon, so boasted by Grecian and Roman epicures." "It is probable that the inhabitants of Lebanon have made no change in their ancient method of making wines."—*Bacchus*, p. 374, note.

Dr. Bowring, in his report on the commerce of Syria, praises, as of excellent quality, a wine of Lebanon consumed in some of the convents of Lebanon, known by the name of vino d'or—golden wine. (Is this the yellow wine which Volney says is too sweet and sugary?) But the Doctor adds "that the habit of boiling wine is almost universal."—*Kitto*, ii. 956.

Caspar Neuman, M.D., Professor of Chemistry, Berlin, 1759, says: "It is observable that when sweet juices are boiled down to a thick consistence, they not only do not ferment in that state, but are not easily brought into fermentation when diluted with as much water as they had lost in the evaporation, or even with the very individual water that exhaled from them."—*Nott*, Lond. Ed., p. 81.

Adams' *Roman Antiquities*, first published in Edinburgh, 1791, on the authority of Pliny and Virgil, says: "In order to make wine keep, they used to boil (deconquere) the must down to one-half, when it was called defrutum, to one-third, sapa."

Smith's *Greek and Roman Antiquities:* "A considerable quantity of must from the best and oldest vines was inspissated by boiling, being then distinguished by the Greeks under the general name Epsuma or Gleuxis, while the Latin writers have various terms, according to the extent to which the evaporation was carried; as Carenum, one-third; defrutum, one-half; and sapa, two-thirds." Professor Anthon, in his *Greek and Roman Antiquities*, makes the same statement.

Cyrus Reading, in his *History of Modern Wines*, says: "On Mount Lebanon, at Kesroan, good wines are made, but they are for the most part *vins cuit* (boiled wines). The wine is preserved in jars."—*Kitto*, ii. 956.

Dr. A. Russell, in his *Natural History of Aleppo*, considers its wine (Helbon) to have been a species of sapa. He says: "The inspissated juice of the grape, sapa vina, called here *dibbs*, is brought to the city in skins and sold in the public markets; it has much the appearance of coarse honey, is of a sweet taste, and in great use among the people of all sorts."—*Kitto*, ii. 956.

Leiber, who visited Crete in 1817, says: "When the Venetians were masters of the island, great quantities of wine were produced at Rettimo and Candia, and it was made by boiling in large coppers, as I myself observed."—*Nott*.

Mr. Robert Alsop, a minister among the Society of Friends, in a letter to Dr. F. R. Lees in 1861, says: "The syrup of grape-juice is an article of domestic manufacture in most every house in the vine districts of the south of France. It is simply the juice of the grape boiled down to the consistence of treacle."—*Bible Com.*, p. xxxiv.

Rev. Dr. Eli Smith, American missionary in Syria, in the *Bibliotheca Sacra* for November, 1846, describes the methods of making wine in Mount Lebanon as numerous, but reduces them to three classes: 1. The simple juice of the grape is fermented. 2. The juice of the grape is boiled down before fermentation. 3. The grapes are partially dried in the sun before being pressed. With characteristic candor, he states that he "had very little to do with wines all his life, and that his knowledge on the subject was very vague until he entered upon the present investigation for the purpose of writing the article." He further as candidly confesses that the " statements contained in his article are not full in every point;" that "it was written in a country where it was very difficult to obtain authentic and exact information." Of the vineyards, he further states that in "an unbroken space, about two miles long by half a mile wide, only a few gallons of intoxicating wine are made. The wine made is an item of no consideration; it is not the most important, but rather the least so, of all the objects for which the vine is cultivated." He also states that " the only form in which the unfermented juice of the grape is preserved is that of dibbs, which may be called *grape-molasses.*" Dr. E. Smith here confirms the ancient usage of boiling the unfermented juice of the grape. The ancients called it wine; the present inhabitants call it dibbs; and Dr. E. Smith calls it grape-molasses. It is the same thing under these various designations. "A rose may smell as sweet by any other name."

The Rev. Henry Holmes, American missionary to Constantinople, in the *Bibliotheca Sacra* for May, 1848, gives the result of his observation. He wrote two years subsequently to Dr. Eli Smith, and has supplied what was lacking in Dr. E. Smith's statements which were "not full on every point." He did not rely upon information from others, but personally examined for himself, and in every case obtained exact and authentic knowledge. He says: "Simple grape-juice, without the addition of any earth to neutralize the acidity, is boiled from four to five hours, so as to reduce it ONE-FOURTH the quantity put in. After the boiling, for preserving it cool, and that it be *less liable to ferment*, it is put into earthen instead of wooden vessels, closely tied over with skin to exclude the air. It ordinarily has not a particle of intoxicating quality, being used freely by both Mohammedans and Christians. Some which I have had on hand for two years has undergone no change." "The manner of making and preserving this unfermented grape-liquor seems to correspond with the receipts and descriptions of certain drinks included by some of the ancients under the appellation of wine."

"The fabricating of an intoxicating liquor *was never the chief* object for which the grape was cultivated among the Jews. Joined with bread, fruits, and the olive-tree, the three might well be representatives of the productions most essential to them, at the same time that they were the most abundantly provided for the support of life." He mentions sixteen uses of the grape, *wine-making* being the *least important*. "I have asked Christians from Diarbekir, Aintab, and other places in the interior of Asia Minor, and all concur in the same statement."

Dr. Eli Smith, as above, testifies that "wine is not the most important, but *the least*, of all the objects for which the vine is cultivated." These statements are fully confirmed by the Rev. Smylie Robson, a missionary to the Jews of Syria, who travelled extensively in the mountains in Lebanon, as may be seen by his letters from Damascus and published in the *Irish Presbyterian Missionary Herald* of April and May, 1845.

The Rev. Dr. Jacobus, commenting on the wine made by Christ, says: "This wine was not that fermented liquor which passes now under that name. All who know of the wines then used will understand rather the unfermented juice of the grape. The present wines of Jerusalem and Lebanon, as we tasted them, were commonly boiled and sweet, without intoxicating qualities, such as we here get in liquors called wines. The boiling prevents the fermentation. Those were esteemed the best wines which were least strong."

The ancients had a motive for boiling the unfermented juice. They knew from experience that the juice, by reason of the heat of the climate and the sweetness of the grapes, would speedily turn sour. To preserve it sweet, they naturally resorted to the simple and easy method of boiling.

The art of distillation was then unknown; it was not discovered till the ninth century.

FILTRATION.

By filtration, the gluten or yeast is separated from the juice of the grape. Whilst the juice will pass through the filtering implements, the gluten will not, and, being thus separated, the necessary conditions of fermentation are destroyed.

Donavan, already quoted, states that, " if the juice be filtered and deprived of its gluten or ferment, the production of alcohol is impossible." Dr. Ure says, as previously stated, that fermentation may be prevented "by the separation of the yeast either by the filter or by subsidence."

The ancient writers, when speaking of the removal of the *vim, vi, vires*, that is, the potency or fermentable power of the wine, use the following strong words: eunuchrum, castratum, effœminatum—thus expressing the thoroughness of the process by which all fermentation was destroyed.—*A.-B.* 224. Plutarch, born A.D. 60, in his *Symposium,* says: "Wine is rendered old or feeble in strength when it is frequently filtered. The strength or spirit being thus excluded, the wine neither inflames the brain nor infests the mind and the passions, and is much more pleasant to drink."—*Bible Com.* p. 278. In this passage, we are instructed that the filter was not a mere strainer, such as the milkmaid uses, but was such an instrument as forced the elements of the grape-juice asunder, separating the gluten, and thus taking away the strength, the spirit, which inflames the head and infests the passions.

Pliny, liber xxiii. cap. 24, says: "Utilissimum (vinum) omnibus sacco viribus fractis. The most useful wine has all its force or strength broken by the filter."—*Bible Commentary,* pp. 168 and 211.

Others hold that the true rendering is: "For all the sick, the wine is most useful when its forces have been broken by the strainer." This does not relieve the difficulty; for, when the *forces* of the wine, which is the *alcohol,* have been broken (fractis, from frango, to break in pieces, to dash to pieces), what then is left but the pure juice? The next sentence of Pliny clearly states that the *vires* or forces of the wine are produced by fermentation: "Meminerimus succum esse qui fervendo vires e musto sibi fecerit." "We must bear in mind that there is a *succus,* which, by fermenting, would make to itself a *vires* out of the must." The succus represents the gluten or yeast, the detention of which in the filter would effectually prevent all fermentation.—*Nott,* Edition by F. R. Lees, p. 211. The strainer (saccus) separates the gluten; for in no other way can it break the forces, the fermenting power. Smith, in his *Greek and Roman Antiquities,* says: "The use of the saccus (filter), it was believed, diminished the strength of the liquor. For this reason it was employed by the dissipated in order that they might be able to swallow a greater quantity without becoming intoxicated." Again: "A great quantity of sweet wines was manufactured by checking the fermentation." Prof. C. Anthon makes a similar statement in his *Dictionary of Greek and Roman Antiquities.*

Again, *Pliny:* "Inveterari vina saccisque castrari." " Wines were rendered old and castrated or deprived of all their vigor by filtering."—*Nott*, London Ed.

"Ut plus capiamus vini sacco frangimur vires;" that we may drink the more wine, we break in pieces, vires, the strength or spirit, sacco, by the filter. He adds that they practised various incentives to increase their thirst. —*Bible Commentary,* p. 168.

On the words of Horace, "vina liques," *Car.* lib. i. ode ii., the *Delphin Notes* says: " Be careful to prepare for yourself wine percolated and defecated by the filter, and thus rendered sweet and more in accordance to nature and a female taste." Again: " The ancients filtered and defecated their must repeatedly before they could have fermented; and thus the fæces which nourish the strength of the wine being taken away, they rendered the wine itself more liquid, weaker, lighter and sweeter, and more pleasant to drink."—*Bible Commentary,* p. 168, and *Nott,* London Edition, p. 79.

Captain Treat, in 1845, wrote: "When on the south coast of Italy, last Christmas, I enquired particularly about the wines in common use, and found that those esteemed the best were sweet and unintoxicating. The boiled juice of the grape is in common use in Sicily. The Calabrians keep their intoxicating and unintoxicating wines in separate apartments. The bottles were generally. marked. From enquiries, I found that unfermented wines were esteemed the most. It was drunk mixed with water. Great pains were taken in the vintage season to have a good stock of it laid by. The grape-juice was filtered two or three times, and then bottled, and some put in casks and buried in the earth—some kept in water (to prevent fermentation).—*Dr. Lees' Works*, vol. ii. p. 144.

Gluten is as indispensable to fermentation, whether vinous or acetous, as is sugar. It is a most insoluble body until it comes in contact with the oxygen of the atmosphere; but by frequent filtering of the newly-pressed juice, the gluten is separated from the juice, and thus fermentation prevented.

SUBSIDENCE.

Chemical science teaches that the gluten may be so effectually separated from the juice by subsidence as to prevent fermentation. The gluten, being heavier than the juice, will settle to the bottom by its own weight if the mass can be kept from fermentation for a limited period. Chemistry tells us that, if the juice is kept at a temperature below 45°, it will not ferment. The juice being kept cool, the gluten will settle to the bottom, and the juice, thus deprived of the gluten, cannot ferment. Dr. Ure says: "By lowering the temperature to 45°, if the fermenting mass becomes clear at this temperature and be drawn off from the subsided yeast, it will not ferment again, though it should be heated to the proper pitch."—*Bible Commentary*, p. 168.

Pliny, liber xiv. c. 9, when speaking of a wine called

Aigleuces, that is, always sweet, says: "Id evenit cura." "That wine is produced by care." He then gives the method: "Mergunt eam protinus in aqua cados donec bruma transeat et consuetudo fiat algendi." " They plunge the casks, immediately after they are filled from the vat, into water, until winter has passed away and the wine has acquired the habit of being cold."—*Kitto*, ii. 955; *A.-B.* 217; *Smith's Antiquities*. Being kept below 45°, the gluten settled to the bottom, and thus fermentation was prevented.

Columella gives the receipt: " Vinum dulce sic facere oportet." " Gather the grapes and expose them for three days to the sun; on the fourth, at mid-day, tread them; take the mustum lixivium ; that is, the juice which flows into the lake before you use the press, and, when it has *settled*, add one ounce of powdered iris; *strain* the wine from its fæces, and pour it into a vessel. This wine will be sweet, firm or durable, and healthy to the body."— *Nott*, London Ed. 213; *A.-B.* 216.

We notice in this receipt: 1, the lixivium, which the lexicon (Leverett) defines "*must, which flows spontaneously from grapes before they are pressed ;*" 2, this is allowed to *settle*, the gluten thus subsiding; 3, pounded iris is put into the juice, and then it is strained or filtered. Here are three combined operations to prevent fermentation.

The same author, liber xii. cap. 29 (see *Nott* and *A.B.* 216), mentions a receipt: " That your must may always be as sweet as when it is new, thus proceed: Before you apply the press to the fruit, take the newest must from the lake, put into a *new amphora*, bung it up, and cover it very carefully with pitch, lest any water should enter; then immerse it in a cistern or pond of pure cold water, and allow no part of the amphora to remain above the surface. After forty days, take it out, and it will remain sweet for a year." Prof. C. Anthon gives the same receipt in his *Dictionary of Greek and Roman Antiquities*. We here notice: 1, that the newest—the unfermented juice—is taken; 2, it is put in a *new* amphora or jar free from all ferment from former use; 3, the air is perfectly excluded; 4, it is immersed in cold water for forty days. Being below 45°, fermentation could not commence. Thus there was ample time for the gluten to settle at the bottom, thus leaving the juice pure and sweet.

Columella, liber xii. cap. 51, gives a receipt for making oleum gleucinum: "To about ninety pints of the best must in a barrel, eighty pounds of oil are to be added, and a small bag of spices sunk to the place where the oil and wine meet; the oil to be poured off on the ninth day. The spices in the bag are to be pounded and replaced, filling up the cask with another eighty pounds of oil; this oil to be drawn off after seven days."—*Bible Commentary*, p. 297. Here notice: 1, The best must—the unfermented juice—is taken; 2, This, when in the cask, is covered with oil, which excludes the air from the juice; 3, A bag of spices is placed in contact with the juice; 4, After nine days, in which the gluten would settle, the oil is poured off; 5, The spices are pounded and replaced, oil again is poured in, to remain seven days, and then drawn off, leaving the juice pure and unfermented.

The ancients preserved some of their wines by *depurating* them. "The must, or new wine," says Mr. T. S. Carr, " was refined with the yolks of pigeon eggs (*Roman Antiquities*), which occasioned the subsidence of the albumen or ferment. But on the new wine being allowed to stand, this principle would subside by natural gravity; hence the ancients poured off the upper and luscious portion of the wine into another vessel, repeating the process as often as necessary, until they procured a clear, sweet wine which would keep."—*Kitto*, ii. 955.

Harmer, on the authority of Charden, observes that "in the East they frequently pour wine from vessel to vessel; for when they begin one, they are obliged immediately to empty it into smaller vessels or into bottles, or it would grow sour." Chemistry teaches that sweet juices in hot climates, if left to themselves, immediately pass into the acetous fermentation and become sour. To avoid this the above process was adopted.

FUMIGATION.

Dr. Ure states that fermentation may be stopped by the application or admixture of substances containing sulphur; that the operation consists partly in absorbing oxygen, whereby the elimination of the yeasty particles is prevented. Adams in his *Roman Antiquities*, on the authority of Pliny and others, says "that the Romans fumigated their wines with the fumes of sulphur; that they also mixed with the mustum, newly pressed juice, yolks of eggs, and other articles containing sulphur. When thus defæcabantur (from defæco, 'to cleanse from the dregs, to strain through a strainer, refine, purify, defecate'), it was poured (diffusum) into smaller vessels or casks covered over with pitch, and bunged or stopped up."

Gardiner, in his *Dictionary of the Arts*, article Wine, says: "The way to preserve *new wine*, in the state of *must*, is to put it up in very strong but small casks, firmly closed on all sides, by which means it will be kept from fermenting. But if it should happen to fall into fermentation, the only way to stop it is *by the fumes of sulphur.*"—*Thayer*, p. 22.

Here we notice two important facts. The first is, that the exclusion of the air from the fresh juice will prevent fermentation. The second is, that, when fermentation has commenced, the fumes of sulphur will arrest it. How more certainly it will prevent fermentation if applied to the new wine.

Cyrus Reading says of sulphur, "Its object is to impart to wine clearness and the principle of preservation, and to prevent fermentation."—*Nott*, London Ed. p. 82.

Mr. T. S. Carr says that the application of the fumarium to the mellowing of wines was borrowed from the Asiatics, and that the exhalation would go on until the wine was reduced to the state of a syrup."—*Kitto*, ii. 956.

"Such preparations," says Sir Edward Barry, "are made by the modern Turks, which they frequently carry with them on long journeys, and occasionally take as a strengthening and reviving cordial."—*Kitto*, ii. 956.

"In the *London Encyclopædia* ' stum ' is termed an unfermented wine; to prevent it from fermenting, the casks are matched, or have brimstone burnt in them."—*Nott*, London Ed. p. 82.

Count Dandalo, on the *Art of Preserving the Wines of Italy*, first published at Milan, 1812, says, "The last process in wine-making is sulphurization: its object is to secure the most long-continued preservation of all wines, even of the very commonest sort."—*Nott*.

A familiar illustration and confirmation may be had from the expressed juice of the apple. If the fresh unfermented apple-juice is not cider, what is it? Every boy, straw in hand, knows that it is cider—so does every farmer and housewife. After it has fermented, it is also called cider. It is a generic word, applicable to the juice of the apple in all its stages, just as *yayin* in the Hebrew, *oinos* in the Greek, *vinum* in the Latin, and *wine* in English are generic words, and denote the juice of the grape in all conditions. When the barrel is filled with the fresh unfermented juice of the apple, add sulphur, or mustard-seed, make the barrel air-tight, and keep it where it is cold, and fermentation will not take place. When the gluten has subsided and, by its specific gravity, has settled at the bottom, the pure unfermented juice may be bottled and kept sweet. This, men call cider; they have no other name for it.

In all these four methods, but one object is sought—it is to preserve the juice sweet.

DID THE ANCIENTS USE AND CALL THEM WINE?

In all the extracts we have made in the preceding pages, the writers call the grape-juice *wine*, whether boiled or filtered, or subsided or fumigated. It may be well again to refer to a few cases.

Pliny says the "Roman wines were as thick as honey," also that the "Albanian wine was very sweet or luscious, and that it took the third rank among all the wines." He also tells of a Spanish wine in his day, called "inerticulum"—that is, would not intoxicate—from "iners," inert, without force or spirit, more properly termed "justicus sobriani,". sober wine, which would not inebriate. —*Anti-Bac.* p. 221.

According to Plautus, B.C. 200, even *mustum* signified both wine and sweet wine.—*Nott*, London Ed. p. 78.

Nicander says: "And Æneus, having squeezed the

juice into hollow cups, called it wine (oinon)."—*Nott*, p. 78. " The Greeks as well as the Hebrews called the fresh juice wine."—*Nott*, London Ed. p. 78.

Columella says the Greeks called this unintoxicating wine " Amethyston," from Alpha, negative, and methusis, intoxicate—that is, a wine which would not intoxicate. He adds that it was a *good wine*, harmless, and called " iners," because it would not affect the nerves, but at the same time it was not deficient in flavor.—*A.-B.* p. 221.

Aristotle says of sweet wine, glukus, that it would not intoxicate. And that the wine of Arcadia was so thick that it was necessary to scrape it from the skin bottles in which it was contained, and dissolve the scrapings in water.—*Nott*, London Ed. p. 80.

Homer (*Odyssey*, book ix.) tells us that Ulysses took in his boat " a goat-skin of sweet black wine, a divine drink, which Marion, the priest of Apollo, had given him—it was sweet as honey—it was imperishable, or would keep for ever; that when it was drunk, it was diluted with twenty parts water, and that from it a sweet and divine odor exhaled."—*Nott*, London Ed. p. 55.

Horace, liber i. ode xviii. line 21, thus wrote:

> " Hic innocentis pocula Lesbii
> Duces sub umbra."

Professor Christopher Smart, of Pembroke College, Cambridge, England, more than a hundred years since, when there was no controversy about fermented or unfermented wines, thus translated this passage: " Here shall you quaff, under a shade, cups of *unintoxicating wine*."

Again, we read in Horace, liber. iii. ode viii. line 9:

"Hic dies, anno redeunte, festus,
Corticem adstrictum pice divomebit
Amphoræ fumum bibere institutæ
Consule Tullo.

"Sume, Mæcenas, cyathos amici
Sospitis centum; et vigiles lucernas
Perfor in lucem: procul omnis esto
Clamor et ira."

I take again the translation of Professor Smart: "This day, sacred in the revolving year, shall remove the cork fastened with pitch from that jar which was set to fumigate in the consulship of Tullus. Take, my Mæcenas, an hundred glasses, on account of the safety of your friend, and continue the wakeful lamps even to daylight: all clamor and passion be far away."

This Horace calls wine—it was *fumigated*—the amphora was corked and fastened with pitch, and that an hundred glasses might be drunk without clamor or passion. The *Delphin Notes* to Horace state, "The ancients filtered their wines repeatedly, before they could have fermented."

Athenæus says: "The sweet wine (*glukus*), which among the Sicilians is called Pollian, may be the same as the *Biblinos oinos*." "Sweet kinds of wines (*oinos*) do not make the head heavy," as Hippocrates says. His words are, "*Glukus* is less calculated than other wine (*oinodeos*) to make the head heavy, and it takes less hold of the mind." He speaks of the mild Chian and the sweet Bibline, and Plautus of the toothless Thanium and Coan, all of which are comprehended under oinos, wine.—*Nott*, London Ed. p. 80.

Professor M. Stuart, on pages 44 and 45 of his *Letter to*

Dr. Nott, published 1848, mentions that some forty years ago Judge Swift told him that, when the Hon. O. Elsworth, the first Chief-Justice of the U. S. Supreme Court, was on his way to France as ambassador, accompanied by Judge Swift, of Connecticut, as secretary, they were shipwrecked and cast upon the coast of Spain. On their way to Paris, among the mountains of Spain, a wine was strongly urged upon them which would not intoxicate. Judge Swift first made the experiment on himself. He found that it did not produce any tendency of the kind. The Chief-Justice and himself used to drink a bottle each with their dinner, and a small bottle at night. It was found to be a precious balm to the ambassador, who had become fearfully exhausted by continued sea-sickness.

Judge Swift, continues the Professor, assured me that "he never, before or since, tasted of anything that would bear comparison with the delicacy and exquisite flavor and refreshing effect of this wine, when taken with due preparation of cooling and mixing with water. He expressed his confident belief that a gallon of it drunk at a time, if a man could swallow down so much, would not affect his head in the least degree."

Polybius states that "among the Romans the women were allowed to drink a wine which is called passum, made from raisins, which drink very much resembled *Aegosthenian and Cretan gleukos* (sweet wine), and which men use for the purpose of allaying excessive thirst."—*Nott*, London Ed. p. 80.

Henderson, in his *History of Wines*, p. 44, commenting on the boiled wine of the Roman women referred to by Virgil (*Georg.* i. 293), truly says, "The use of this inspissated juice became general." Rev. W. H. Rule, in his *Brief Enquiry*, confesses that it was the *protropos* or *prodromos oinos* of the Greeks."—*Nott*, London Ed., Lees' Appendix, p. 221.

Smith's *Greek and Roman Antiquities* says: "That which flowed from the clusters, in consequence of their pressure upon each other, to which the inhabitants of Mytelene gave the name of *protropos*."

The prohibition of intoxicating wines to women was enforced by the severest penalties. "Plato, Aristotle, Plutarch, and others have noticed the hereditary transmission of intemperate propensities, and the legislation that imposed abstinence upon women had unquestionably in view the greater vigor of the offspring—the 'mens sana in corpore sano' (healthy minds in a healthy body)."
—*Bible Commentary*, p. 72.

"Modern medical enquiries have made clear the fact, surmised by some ancient philosophers, of the powerful influence of maternal regimen on the uterine condition and future health of children." "That indulgence in the use of strong drink by expectant mothers would be injurious to their offspring was known to the learned and wise among the ancients."—*Bible Commentary*, p. 72.

Matthew Henry, in the case of Samson, remarks, "Women with child ought conscientiously to avoid whatever they have reason to think will be in any way prejudicial to the health or good condition of the fruit of their body. And perhaps Samson's mother was to refrain from wine and strong drink, not only because he was designed for a Nazarite, but because he was designed for a man of strength, which his mother's temperance would contribute to."

That old Roman *prohibitory law*, which forbade intoxicating wine whilst it allowed the pure juice, was founded in common sense and benevolence. It is to be regretted that they were not as wise and merciful towards themselves as they were towards their wives and the health and strength of their offspring.

Dr. Laurie, who holds that "it is the nature of wine to be fermented," and "that fermentation is essential to its becoming wine," still admits that there are "traces of unfermented wine in classical authors," and that it "is known in history;" which he thus strangely qualifies—known in history "only as one of the unnatural and rare luxuries of the most corrupt period of the Roman Empire." Queer logic this, that unintoxicating wine should indicate the most corrupt period of the Roman Empire! Human nature must have greatly changed, for now the course of history is *rum, rags, ruin*. And experience teaches that the use of intoxicating drinks is associated with desecrated Sabbaths, loose views of morality and religion, and the increase of pauperism, crime, and taxation.

The Rev. W. H. Rule, already named, says: "This very grape-juice, notwithstanding its purity, was chiefly known in antiquity as the *casual drink of the peasantry*, or, when carefully preserved, as *the choice beverage of epicures*. It was sweet to the taste, and had not acquired the asperity consequent on the abstraction (conversion) of saccharine matter by fermentation."—*Nott*, London Ed., Appendix C, p. 222.

Smith, in his *Greek and Roman Antiquities*, says: "The sweet, unfermented juice of the grape was termed *gleukos* by the Greeks and *mustum* by the Romans—the latter word being properly an adjective signifying new or fresh." "A portion of the must was used at once, being drunk fresh." "When it was desired to preserve a quantity in the sweet state, an amphora was taken and coated with pitch within and without, it was filled with mustum lixivium, and corked so as to be perfectly air-tight. It was then immersed in a tank of cold fresh water, or buried in wet sand, and allowed to remain for six weeks or two months. The contents, after this process, was found to remain unchanged for a year, and hence the name, aeigleukos—that is, 'semper mustum,' always sweet."

Chas. Anthon, LL.D., in his *Dictionary of Greek and Roman Antiquities*, gives the same receipt and definitions, and fully sustains the position that these preparations of the unfermented grape-juice were by the ancients known as wine.

We have a great variety of ancient receipts for making different kinds of wine. . Some of them, as we have seen, were not fermented, and therefore not intoxicating. Others were intoxicating. The receipts mention the different articles out of which wines were made, such as millet, dates, lotus-tree, figs, beans, pears, pomegranates, myrtle, hellebore, etc. Foreign ingredients were formerly added to wines to make them intoxicating. These wines were not approved, and towards these not temperance but total abstinence was enjoined. Various drugs are specified by which the juice was made more potent, as wormwood, hellebore. We learn from Homer that Helen prepared for Telemachus a cup in which a powerful drug was infused. Also, that Circe made use of "direful drugs." Such preparations were common in the East. The Orientals of the present day have a knowledge of drugs which they combine with beverages for profligate purposes. We read in Isaiah v. 22 of "men of strength to mingle strong drink." The juice of the grape was "mixed with pungent and heady drugs in order to gratify a base and insatiable appetite." Particularly, in Lamentations of Jeremiah

iii. 15 we read, "He hath made me drunk with wormwood." J. G. Koht, in his *Travels in Austria*, mentions a wine of wormwood. To make it, the juice is boiled with certain herbs. This wine decoction is as renowned in Hungary as the Tokay Essence.—*Bible Commentary*, p 203.

The divine anger is symbolized by the cup which is "full of mixture;" Ps. lxxv. 8; "cup of his fury," Isaiah li. 17; "wine-cup of his fury," Jer. xxv. 15.

We cannot imagine that Pliny, Columella, Varro, Cato, and others were either cooks or writers of cook-books, but were intelligent gentlemen moving in the best circles of society. So when they, with minute care, give the receipts for making sweet wine, which will remain so during the year, and the processes were such as to prevent fermentation, we are persuaded that these were esteemed in their day. That they were so natural and so simple as to like these sweet, harmless beverages is rather in their favor, and not to be set down against them. That there were men in their day, as there are many in ours, who loved and used intoxicating drinks, is a fact which marked their degradation.

WINE WITH WATER.

There is abundance of evidence that the ancients mixed their wines with water; not because they were so strong, with alcohol, as to require dilution, but because, being rich syrups, they needed water to prepare them for drinking. The quantity of water was regulated by the richness of the wine and the time of year.

"Those ancient authors who treat upon domestic manners abound with allusions to this usage. Hot water, tepid water, or cold water was used for the dilution of wine according to the season." "Hesiod prescribed, during the summer months, three parts of water to one of wine." "Nicochares considers two parts of wine to five of water as the proper proportion." "According to Homer, Pramnian and Meronian wines required twenty parts of water to one of wine. Hippocrates considered twenty parts of water to one of the Thracian wine to be the proper beverage." "Theophrastus says the wine at Thasos is wonderfully delicious." Athenæus states that the Tæniotic has such a degree of richness or fatness that when mixed with water it seemed gradually to be diluted, much in the same way as Attic honey well mixed.—*Bible Commentary*, p. 17.

Captain Treat says, " The unfermented wine is esteemed the most in the south of Italy, and wine is drunk mixed with water."—*Lees' Works*. Also in Spain and Syria.

"In Italy the habit (mixing wine with water) was so universal that there was an establishment at Rome for the public use. It was called THERMOPOLIUM, and, from the accounts left of it, was upon a large scale. The remains of several have been discovered among the ruins of Pompeii. Cold, warm, and tepid water was procurable at these establishments, as well as wine, and the inhabitants resorted there for the purpose of drinking, and also sent their servants for hot water."—*Nott*, London Ed. p. 83.

" The annexed engraving of the THERMOPOLIUM is copied from the scarce work of Andreas Baccius (*De Nat. Vinorum Hist.*, Rome, 1597, lib. iv. p. 178). The plan was obtained by himself, assisted by two antiquaries, from the ruins of the Diocletian Baths (Rome). Nothing can more clearly exhibit the contrast between the ancient wines and those of modern Europe than the widely different mode of treating them. The hot water was often necessary, says Sir Edward Barry, to dissolve their more inspissated and old wines."—*Kitto*, ii. p. 956.

"Nor was it peculiar to pagans to mingle water with wine for beverage and at feasts; nor to profane writers to record the fact. It *is* written of Wisdom, she mingled her wine—Prov. ix. 2—and so written by an inspired penman."—*Nott*, London Ed. p. 84.

This mixed wine must be different from that named in Ps. lxxv. 8. "full of mixture," which we have seen is the symbol of the divine vengeance, the cup prepared for his enemies. But in Prov. ix. 2, it is a blessing to which friends are invited. If in this passage the mixture is of aromatic spices, in addition to the water necessary to dilute the syrup, it was not to fire the blood with alcohol, but to gratify the taste with delicate flavors.

The Passover was celebrated with wine mixed with water. According to Lightfoot, each person—man, woman, and child—drank four cups. Christ and his disciples having celebrated the Passover, he took of the bread and the wine that remained, and instituted the Lord's Supper. The wine was, as we believe, the rich syrup diluted with water. This kind of wine met all the requirements of the law concerning leaven—the true rendering of *Matsah*, according to Dr. D. F. Lees, being *unfermented things*. The conclusion to which these varied sources of proof bring us may thus be stated:

1. That unfermented beverages existed, and were a common drink among the ancients.
2. That to preserve their very sweet juices, in their hot climate, they resorted to boiling and other methods which destroyed the power and activity of the gluten, or effectually separated it from the juice of the grape.
3. That these were called wines, were used, and were highly esteemed.

Prof. M. Stuart says, " Facts show that the ancients not only preserved their wine unfermented, but regarded it as of a higher flavor and finer quality than fermented wine."—*Letter to Dr. Nott*.

That they also had drinks that would intoxicate cannot be denied. All that we have aimed to show is that intoxicating wines were *not the only wines in use*.

With the teachings of chemical science, and with the knowledge of the tastes and usages of the ancients, we are the better prepared to examine and understand the Bible, which was written when those tastes and usages were in actual operation. Common honesty demands that we interpret the Scriptures with the eye, the taste, and the usages of the ancients, and not with the eye, the taste, and the usages of the moderns. We should interpret each text so as to be in harmony not only with the drift and scope of the whole teachings of the Bible, but also with the well-ascertained and established laws of nature. It certainly is as important to harmonize the interpretations of the Bible with the teachings of chemistry and the laws of our physical, intellectual, and moral nature, violated by alcoholic drinks, as it is to harmonize the interpretations of the same word of God with the ascertained facts of geology and astronomy. To these latter topics, Biblical scholars have given most praiseworthy attention. Let the same anxious interest animate our endeavors to harmonize the Bible teachings with clearly ascertained facts and with the truth which the temperance reformation has made indisputable.

The will of God registered in the laws of nature, and the will of God registered in the inspired revelation, cannot possibly contradict each other. They must harmonize. Whatever difficulties may now stand in the way of this harmony, we know that, as science becomes more intelligently informed of the laws of nature, and as the interpretation of the Bible becomes more thorough and emancipated, the testimony of God's works and word will perfectly harmonize.

"The books of nature and revelation were written by the same unerring hand. The former is more full and explicit in relation to the physical, the latter in relation to the moral, laws of our nature; still, however, where both touch on the same subject, they will ever be found, *when rightly interpreted*, to be in harmony." "*Nature and revelation are as little at variance on the wine question as on other questions, and when rightly consulted it will be found to be so.* It is not in the *text*, but in the *interpretation*, that men have felt straitened in their consciences; and though this feeling should continue, unless the providence of God changes, it will not alter the facts of the case."—*Nott*, London Ed. p. 75.

THE SCRIPTURES.

It should be constantly borne in mind that the Authorized Version was translated when the drinking usages were well-nigh universal. The attention of Christians and of thoughtful men had not been called to the pernicious influence of alcoholic drinks. Though drunkenness existed, still no plans were then devised either for its prevention or its cure. It was regarded as an evil incident to hospitality and social cheer.

The translators, with the most honest purpose, faithfully, according to their ability, rendered into English the original Scriptures, but were nevertheless unintentionally and unconsciously influenced by the philosophy and usages of their day. As the river carries in its waters that which with absolute certainty tells of the soil through which it has flowed, so the translators must carry into the renderings which they give evidences of the prevailing usages and modes of thought of their day. Thus innocently, though naturally, shades of meaning have been given to particular passages. These have come down to us with feelings of sacred reverence. To give a new rendering seems to be almost sacrilege. With this feeling every department of science has to contend when it would throw new light upon the sacred page. Astronomy and geology have met this difficulty, and it is not strange that the cause of temperance should have to contend with this feeling, notwithstanding the convictions of temperance men are the result of experience and diligent, patient investigation.

We would not distrust, much less weaken, confidence in the Word of God. We would, however, remind the reader that ONLY THE ORIGINAL TEXT IS INSPIRED; that no translation, much less no mere human interpretation, is ultimate authority.

GENERIC WORDS.

Professor M. Stuart, in his *Letter to Rev. Dr. Nott*, February 1, 1848, says, page 11 : " There are in the Scriptures (Hebrew) but two *generic words* to designate such drinks as may be of an intoxicating nature when fermented and which are not so before fermentation. In the Hebrew Scriptures the word *yayin*, in its broadest meaning, designates *grape-juice*, or *the liquid which the fruit of the vine yields*. This may be new or old, sweet or sour, fermented or unfermented, intoxicating or unintoxicating. The simple idea of *grape-juice* or *vine-liquor* is the basis and essence of the word, in whatever connection it may stand. The specific sense which we must often assign to the word arises not from the word itself, but from the connection in which it stands."

He justifies this statement by various examples which illustrate the comprehensive character of the word.

In the London edition (1863) of President E. Nott's *Lectures*, with an introduction by Tayler Lewis, LL.D., Professor of Greek in Union College, and several appendices by F. R. Lees, he says : " Yayin is a generic term, and, when not restricted in its meaning by some word or circumstance, comprehends vinous beverage of every sort, however produced. It is, however, as we have seen, *often* restricted to the fruit of the vine in its natural and unintoxicating state" (p. 68).

Kitto's Cyclopædia, article Wine : " Yayin in Bible use is a very general term, including every species of wine made from grapes (vinos ampelinos), though in later ages it became extended in its application to wine made from other substances."

Rev. Dr. Murphy, Professor of Hebrew at Belfast, Ireland, says: " Yayin denotes all stages of the juice of the grape."

" Yayin (sometimes written yin, yain, or ain) stands for the expressed juice of the grape—the context sometimes indicating whether the juice had undergone or not the process of fermentation. It is mentioned one hundred and forty-one times."—*Bible Commentary*, Appendix B, p. 412.

SHAKAR, " the second, is of the like tenor," says Professor Stuart, page 14, but applies wholly to a different liquor. The Hebrew name is *shakar*, which is usually translated *strong drink* in the Old Testament and in the New. The mere English reader, of course, invariably gets from this translation a wrong idea of the real meaning of the original Hebrew. He attaches to it the idea which the English phrase now conveys among us, viz., that of a *strong, intoxicating drink*, like our *distilled* liquors. . As to *distillation*, by which alcoholic liquors are now principally obtained, it was utterly unknown to the Hebrews, and, indeed to all the world in ancient times." " The true original idea of *shakar* is a *liquor obtained from dates or other fruits* (grapes excepted), or barley, *millet*, etc., which were dried, or scorched. and a decoction of them was mixed with honey, aromatics, etc."

On page 15 he adds: "Both words are *generic*. The first means vinous liquor of any and every kind; the second means a corresponding liquor from dates and other fruits, or from several grains. Both of the liquors have in them the *saccharine principle ;* and therefore they may become alcoholic. But both may be kept and used in an *unfermented* state; when, of course, no quantity that a man could drink of them would intoxicate him in any perceptible degree." "The two words which I have thus endeavored to define are the *only two* in the Old Testament which are *generic*, and which have reference to the subject now in question."

"SHAKAR (sometimes written shechar, shekar) signifies 'sweet drink' expressed from fruits other than the grape, and drunk in an unfermented or fermented state. It occurs in the O. T. twenty-three times."—*Bible Commentary*, p. 418. *Kitto's Cyclopædia* says: "*Shakar* is a generic term, including palm-wine and other *saccharine* beverages, except those prepared from the vine." It is in this article defined "*sweet drink.*"

Dr. F. R. Lees, page xxxii. of his Preliminary Dissertation to the *Bible Commentary*, says *shakar*, "saccharine drink," is related to the word for sugar in all the Indo-Germanic and Semitic languages, and is still applied throughout the East, from India to Abyssinia, to the palm sap, the *shaggery* made from it, to the date juice and syrup, as well as to sugar and to the fermented palm-wine. It has by usage grown into a generic term for 'drinks,' including fresh juices and inebriating liquors other than those coming from the grape. See under the heading, "Other Hebrew Words" for further illustrations, page 58.

TIROSH, in *Kitto's Cyclopædia*, is defined "vintage fruit." In *Bible Commentary*, p. 414: "Tirosh is a collective name for the natural produce of the vine." Again, *Bible Commentary*, p. xxiv.: "Tirosh is not wine at all, but the fruit of the vineyard in its natural condition." A learned Biblical scholar, in a volume on the wine question, published in London, 1841, holds that tirosh is not wine, but fruit. This doubtless may be its meaning in some passages, but in others it can only mean wine, as, for example, Prov. iii. 10: "Thy presses shall burst out with new wine" (tirosh); Isa. lxii. 8: "The sons of the stranger shall not drink thy new wine" (tirosh).

"On the whole, it seems to me quite clear," says Prof. Stuart, p. 28, "that tirosh is a species of wine, and not a genus, like yayin, which means *grape-juice* in any form, or of any quality, and in any state, and usually is made definite only by the context."

"Tirosh is connected with corn and the fruit of the olive and the orchard nineteen times; with corn alone, eleven times; with the vine, three times; and is otherwise named five times: in all, thirty-eight times." "It is translated in the Authorized Version twenty-six times by wine, eleven times by new wine (Neh. x. 39, xiii. 5, 12; Prov. iii. 10; Isa. xxiv. 7, lxv. 8; Hos. iv. 14, ix. 2; Joel i. 10; Hag. i. 11; Zach. ix. 17), and once (Micah vi. 15) by 'sweet wine,' where the margin has new wine."—*Bible Commentary*, p. 415.

So uniform is the good use of this word that there is but one doubtful exception (Hosea iv. 11): "Whoredom and wine (yayin), and new wine (tirosh), take away the heart." Here are three different things, each of which is charged with taking away the heart. As whoredom is not the same as yayin, so yayin is not the same as tirosh. If physical intoxication is not a necessary attribute of the first, then why is it of the third, especially when the second is adequate for intoxication? If yayin and tirosh each means intoxicating wine, then why use both? It would then read, whoredom and yayin (intoxicating wine) and tirosh (intoxicating wine) take away the heart, which is tautological. The three terms are symbolical.

Whoredom is a common designation of idolatry, which the context particularly names. This steals the heart from God as really as does literal whoredom.

Yayin may represent drunkenness or debased sensuality. This certainly takes away the heart.

Tirosh may represent luxury, and, in this application, dishonesty, as tirosh formed a portion of the tithes, rapacity in exaction, and perversion in their use, is fitly charged with taking away the heart.

Certain interpreters imagine that only alcoholic drinks take away the heart; but we know from the Bible that pride, ambition, worldly pleasures, fulness of bread, Ezek. xvi. 49, and other things, take away the heart.

G. H. Shanks, in his review of Dr. Laurie, says: "In vine-growing lands, grapes are to owners what wheat, corn, flax, etc., are to agriculturists, or what bales of cotton or bank-notes are to merchants. Do these never take away the heart of the possessor from God?"

OTHER HEBREW WORDS.

We extract from Dr. F. R. Lees' Appendix B of *Biblical Commentary* the following, pp. 415–418:

KHEMER is a word descriptive of the foaming appearance of the juice of the grape newly expressed, or when undergoing fermentation. It occurs but nine times in all, including once a verb, and six times in its Chaldee form of *khamar* or *khamrah*. Deut. xxxii. 14; Ezra vi. 9, vii. 22; Ps. lxxv. 8; Isa. xxvii. 2; Dan. v. 1, 2, 4, 23.

Liebig says: "Vegetable juice in general becomes turbid when in contact with the air BEFORE FERMENTATION COMMENCES."—*Chemistry of Agriculture*, 3d edition. " Thus, it appears, *foam* or *turbidness* (what the Hebrews called khemer, and applied to the foaming blood of the grape) is no proof of alcohol being present."—*Bible Commentary*, Prelim. xvi. note.

AHSIS (sometimes written *ausis, asie, osis*) is specially applied to the juice of newly-trodden grapes or other fruit. It occurs five times. Cant. viii. 2; Isa. xlix. 26; Joel i. 5, iii. 18, Amos ix. 13.

SOVEH (sometimes written *sobe, sobhe*) denotes a luscious and probably boiled wine (Latin, sapa). It occurs three times. Isa. i. 22; Hosea iv. 18; Nahum i. 14.

"It is chiefly interesting as affording a link of connection between classical wines and those of Judea, through an obviously common name, being identical with the Greek *hepsema*, the Latin *sapa*, and the modern Italian and French *sabe*—boiled grape-juice. The inspissated wines, called *defrutum* and *syræum*, were, according to Pliny (xiv. 9), a species of it. The last name singularly suggests the instrument in which it was prepared—the *syr*, or caldron."—*Bible Commentary*, Prelim. xxiii.

MESEK (sometimes written *mesech*), literally, a mixture, is used with its related forms, *mezeg* and *mimsak*, to denote some liquid compounded of various ingredients. These words occur as nouns four times, and in a verbal shape five times. Ps. lxxv. 8; Prov. xxiii. 30; Cant.

vii. 2; Isa. lxv. 11. The verbal forms occur Prov. ix. 2, 5; also, in Ps. cii. 9; Isa. xix. 14.

Ashishah (sometimes written *eshishah*) signifies some kind of fruit-cake, probably cake of pressed grapes or raisins. It occurs four times, and in each case is associated by the Authorized Version with some kind of drink. 2 Sam. vi. 19; 1 Chron. xvi. 3; Cant. ii. 5; Hosea iii. 1.

Shemarim is derived from *shamar*, to preserve, and has the general signification of things preserved. It occurs five times. In Exodus xii. 42, the same word, differently pointed, is twice translated as signifying *to be kept* (observed). Ps. lxxv. 8, dregs; Isa. xxv. 6, fat things; Jer. xlviii. 11, lees; Zeph. i. 12, lees.

Mamtaqqim is derived from *mahthuq*, to suck, and denotes sweetness. It is applied to the mouth (Cant. v. 16) as full of sweet things. In Neh. viii. 10, "drink the sweet" mamtaqqim, sweetness, sweet drinks.

Shakar (sometimes written *shechar, shekar*) signifies sweet drink expressed from fruits other than the grape, and drunk in an unfermented or fermented state. It occurs in the Old Testament twenty-three times. Lev. x. 9; Numb. vi. 3 (twice wine and vinegar), xxviii. 7; Deut. xiv. 26, xxix. 6; Judges xiii. 4, 7, 14; 1 Sam. i. 15; Ps. lxix. 12; Prov. xx. 1, xxxi. 4, 6; Isa. v. 11, 22, xxiv. 9, xxviii. 7, xxix. 9, lvi. 12; Micah ii. 11. Shakar is uniformly translated strong drink in the Authorized Version, except in Numb. xxviii. 7 (strong wine), and in Ps. lxix. 12, where, instead of drinkers of *shakar*, the Authorized Version reads *drunkards*. (See "Generic Words.")

GREEK, LATIN, AND ENGLISH GENERIC WORDS.

Oinos.—Biblical scholars are agreed that in the Septuagint or Greek translation of the Old Testament and in the New Testament, the word *oinos* corresponds to the Hebrew word *yayin*. Stuart says: "In the New Testament we have oinos, which corresponds exactly to the Hebrew yayin."

As both yayin and oinos are generic words, they designate the juice of the grape in all its stages.

In the Latin we have the word vinum, which the lexicon gives as equivalent to oinos of the Greek, and is rendered by the English word wine, both being generic. Here, then, are four generic words, *yayin, oinos, vinum,* and *wine*, all expressing the same generic idea, as including all sorts and kinds of the juice of the grape. Wine is generic, just as are the words groceries, hardware, merchandise, fruit, grain, and other words.

Dr. Frederick R. Lees, of England, the author of several learned articles in *Kitto's Cyclopædia*, in which he shows an intimate acquaintance with the ancient languages, says: "In Hebrew, Chaldee, Greek, Syriac, Arabic, Latin, and English, the words for wine in all these languages are *originally*, and always, and *inclusively*, applied to the blood of the grape in its primitive and natural condition, as well, subsequently, as to that juice both boiled and fermented."

Dr. Laurie, on the contrary, says: "This word denotes intoxicating wine in some places of Scripture; therefore, it denotes the same in all places of Scripture." This not only begs the whole question, but is strange, very strange logic. We find the word which denotes the spirit often rendered wind or breath; shall we, therefore, conclude it always means wind or breath, and, with the Sadducees, infer that there is neither angel nor spirit, and that there can be no resurrection? So, also, because the word translated heaven often means the atmosphere, shall we conclude that it always means atmosphere, and that there is no such place as a heaven where the redeemed will be gathered and where is the throne of God?

But the misery and delusion are that most readers of the Bible, knowing of no other than the present wines of commerce, which are intoxicating, leap to the conclusion, wine is wine all the world over—as the wine of our day is inebriating, therefore the wine mentioned in the Bible was intoxicating, and there was none other.

There is a perverse tendency in the human mind to limit a generic word to a particular species.

John Stuart Mill, in his *System of Logic*, says: "A generic term is always liable to become limited to a single species if people have occasion to think and speak of that species oftener than of anything else contained in the genus. The tide of custom first drifts the word on the shore of a particular meaning, then retires and leaves it there."

The truth of this is seen every day in the way in which the readers of the Bible limit the generic word wine to one of the species under it, and that an intoxicating wine.

CLASSIFICATION OF TEXTS.

The careful reader of the Bible will have noticed that in a number of cases wine is simply mentioned, without anything in the context to determine its character. He will have noticed another class, which unmistakably denotes the bad character of the beverage. He will also have noticed a third class, whose character is as clearly designated as good.

It would extend this discussion too much to trace out all the different ways in which the generic word wine is used. It will suffice to direct attention to the two classes which designate their character.

BAD WINE.

One class of texts refers to wine:

1. *As the cause of intoxication.* This is not disputed.
2. *As the cause of violence and woe.* Prov. iv. 17: "They drink the yayin, wine, of violence." Prov. xxiii. 29, 30: "Who hath woe? Who hath sorrows? Who hath contentions? Who hath babbling? Who hath wounds without cause? Who hath redness of eyes? They that tarry long at the *yayin*, wine; they that go to seek mixed wine."
3. *As the cause of self-security and irreligion.* Isa. lvi. 12: "Come ye, say they, I will fetch *yayin*, wine, and we will fill ourselves with strong drink; and to-morrow shall be as this day, and much more abundant." Hab. ii. 5: "Yea also, because he transgresseth by yayin, wine, he is a proud man, neither keepeth at home, who enlargeth his desire as hell, and is as death, and cannot be satisfied." Isa. xxviii. 7: "They also have erred through yayin, wine, and through strong drink are out of the way; the priest and the prophet have erred through strong drink; they err in vision, they stumble in judgment."
4. *As poisonous and destructive.* Prov. xxiii. 31: "Look not thou upon the yayin, wine, when it is red, when it giveth his color in the cup, when it moveth itself aright. At the last it biteth like a serpent, and stingeth like an adder." Chemists find in this passage an admirable description of the process of vinous fermentation by which alcohol is produced.

It is worthy of particular notice that it is this kind of

wine that men are exhorted and warned not even to look upon, much less to drink; and that because its effects will be like the poisonous, deadly bite of a serpent and the equally fatal sting of the adder. Deut. xxxii. 33: " Their yayin, wine, is the poison of dragons, and the cruel venom of asps."

The Hebrew word khamah, here rendered poison, occurs eight times, and is six times translated poison, as in Deut. xxxii. 24: " The *poison* of serpents;" xxxii. 23: " Their wine is the *poison* of dragons;" Ps. lviii. 4: " Their *poison* is like the *poison* of a serpent;" cxl. 3: " Adders' *poison* is under their lips;" Job vi. 4: " The *poison* whereof drinketh up my spirit."

Hosea vii. 5: " Made him sick with bottles of wine" (khamath), poison; margin, "heat through wine." Hab. ii. 15: " Woe unto him that giveth his neighbor drink; that putteth thy bottle to him." The word *bottle* is rendered khamah, which means *poison*, and is so rendered generally; by a figure, the bottle is put for the poison it contained.

Parkhurst defines this word " *an inflammatory poison,*" and refers to the rabbins, who have identified it with the poisoned cup of malediction. Archbishop Newcome, in his translation, says that " khamah is *gall poison.*" St. Jerome's Version has *gall* in one text, and *mad* in another. —*Nott*, London Ed., F. R. Lees, Appendix A, p. 197. Dr. Gill renders the word, " *thy gall, thy poison.*" The late Professor Nordheimer, of the Union Theological Seminary, New York City, in his *Critical Grammar*, has " *maddening wine.*"

Notice the character given to this wine: gall, poison, poison of serpents, adders' poison, poison of dragons, poison which drinketh up the spirits, maddening wine.

How exact the agreement between the declarations of the Bible and the teachings of physical truth! Alcohol is certified by thousands of illustrations as poison to the human system.

No wonder that against such wine the Scriptures lift up their earnest warnings, because wine (yayin) is a mocker; because it "biteth like a serpent, and stingeth like an adder."

5. *As condemning those who are devoted to drink.* Isa. v. 22: "Woe unto them that are mighty to drink (yayin) wine, and men of strength to mingle strong drink: which justify the wicked for reward, and take away the righteousness of the righteous from him! Therefore as the fire devoureth the stubble, and the flame consumeth the chaff, so their root shall be as rottenness, and their blossom shall go up as dust: because they have cast away the law of the Lord of hosts, and despised the word of the Holy One of Israel."

1 Cor. vi. 10: "Nor drunkards shall inherit the kingdom of God."

6. *As the emblem of punishment and of eternal ruin.* Ps. lx. 3: "Thou hast made us to drink the (yayin) wine of astonishment;" literally, "wine of reeling or trembling." The Vulgate reads, "suffering." Ps. lxxv. 8: "For in the hand of the Lord there is a cup, and the (yayin) wine is red; it is full of mixture; and he poureth out of the same: but the dregs thereof, all the wicked of the earth shall wring them out, and drink them." Isa. li. 17: "O Jerusalem, which hast drunk at the hand of the Lord the cup of his fury; thou hast drunken the dregs of the cup of trembling, and wrung them out;" also, verse 22. Jer. xxv. 15: "Take the yayin, wine-cup, of this fury at my hand." Rev. xvi. 19: "To give unto

her the cup of the (oinou) wine of the fierceness of his wrath." Rev. xiv. 10: "The same shall drink of the (oinou) wine of the wrath of God, which is poured out without mixture into the cup of his indignation; and he shall be tormented with fire and brimstone in the presence of the holy angels, and in the presence of the Lamb: and the smoke of their torment ascendeth up for ever and ever."

GOOD WINE.

From this terrible but very imperfect setting forth of the testimonies of the Bible in regard to the wine whose character is bad, I turn, with a sense of grateful pleasure, to another class of texts which speaks with approbation of a wine whose character is good, and which is commended as a real blessing.

1. *This wine is to be presented at the altar as an offering to God.* Numb. xviii. 12: "All the best of the oil, and all the best of the wine, and of the wheat, the first-fruits of them which they shall offer unto the Lord, them have I given thee." In this passage, all the best of the wine (tirosh) is associated with the best of the oil and of the wheat, denoting the most valuable natural productions—the direct gift of God.

That these terms denote the fruit of the soil in their natural state, seems probable from the next verse: "And whatsoever is first ripe in the land, which they shall bring unto the Lord, shall be thine." This was a first fruit-offering. It is associated with oil, and flour, and the first-fruits; it is an "offering of wine for a sweet savor—an offering made by fire, for a sweet savor unto the Lord." Neh. x. 37: "Bring the first-fruits of our dough, and our offerings, and the fruit of all manner of trees, of (tirosh) wine, and of oil," etc. Again, verse 39: "Bring the offering of the corn, of the (tirosh) new wine, and the oil," etc. From these passages, it is held by some that the solid produce of the vineyard was here presented. Chap. xiii. 5: "The tithes of the corn, and (tirosh) new wine; and the oil," etc.; and 13: "The tithe of the corn, and the (tirosh) new wine, and the oil," etc. It is hardly to be credited, when in the law (Levit. ii. 11) all leaven was forbidden as an offering, that God should require a fermented liquor which, of all others, is the most direct cause of wretchedness and woe in this life, and of eternal ruin in the future, as a religious offering; that against the use of which he had uttered his most solemn warnings and denunciations. As all the other articles offered in worship were in their nature pure and harmless—were essential to the comfort and well-being of man, it is passing strange that the wine should be the one exception.

2. *This wine is classed among the blessings, the comforts, the necessaries of life.* When the patriarch Isaac blessed his son Jacob (Gen. xxvii. 28), he said: "Therefore God give thee of the dew of heaven, and the fatness of the earth, and plenty of corn, and (tirosh) wine." The blessing was on the actual growth of the field—that which "the dew and the fatness of the earth produced;" these were the direct gifts of God.

Of this blessing, Isaac afterwards said to Esau (verse 37): "With corn and (tirosh) wine I have sustained him;" that is, I have pledged the divine blessing to secure to him and his posterity in plenty the things necessary for their best comfort and happiness. Therefore we read, Deut. vii. 13: "And he will love thee, and bless thee, and multiply thee; he will also bless the fruit of thy womb, and the fruit of thy land; thy corn, and thy (tirosh) wine, and thine oil; the increase of thy kine and the flocks of thy sheep in the land which he sware unto thy fathers to give thee." The grouping is very significant: the blessing was to rest upon "the fruit of the womb, upon the fruit of the land, which is specified; thy corn, and thy wine, and thine oil; also, the increase of thy kine and flocks of sheep." It is the direct and immediate product of the land. To secure this, God (Deut. xi. 14) promised: "I will give you the rain of your land in his due season, the first rain and the latter rain, that thou mayest gather in thy corn, and thy (tirosh) wine, and thine oil. And I will send grass into thy fields, that thou mayest eat and be full." Prov. iii. 10: "So shall thy barns be filled with plenty, and thy presses shall burst out with (tirosh) new wine."

Albert Barnes, on Isa. xxiv. 7, says: "New wine (tirosh) denotes properly must, or the wine that was newly expressed from the grape and that was not fermented, usually translated new wine or sweet wine."

Isa. lxv. 8: "As the new wine is found in the cluster, and one saith, Destroy it not; for a blessing is in it." Albert Barnes says: "The Hebrew word (tirosh) here used means properly must, or new wine." On the words "for a blessing is in it," he says: "That which is regarded as a blessing, that is, wine." He cites Judges ix. 13 in proof: "Wine which cheereth God and man (tirosh)."

Joel iii. 18: "The mountains shall drop down new wine (tirosh), and the hills shall flow with milk;" *i.e.*, abundance of blessings. These blessed things are the pure, and harmless, and direct products of the land, necessary for the comfort and happiness of man. Is intoxicating wine, which is the emblem of God's wrath and of eternal ruin, among the things blessed? Still further (Ps. civ. 14, 15): "He causeth the grass to grow for the cattle, and herb for the service of man: that he may bring forth food out of the earth; and wine (yayin) that maketh glad the heart of man, and oil to make his face to shine, and bread which strengtheneth man's heart." Again, we read (Judges ix. 13): "And the vine said, Should I leave my (tirosh) wine, which cheereth the heart of God and man?"

Obviously, God can only be cheered or pleased with the fruit of the vine as the product of his own power and the gift of his goodness, and man is cheered with it when he sees the ripening clusters, and when he partakes thereof.

There is a strange impression, very current in our day, that nothing can cheer and exhilarate but alcoholic drinks. Is it not written, Zech. ix. 7, "Corn shall make the young men cheerful, and new wine (tirosh) the maids"? In referring to the nutritious qualities of the corn and wine, the prophet assigns the corn to the young men, and the new wine, *tirosh*, to the maidens. Here the new wine, the must, or unfermented juice, is approbated. Ps. iv. 7: "Thou hast put gladness" (the same word which is translated cheereth in Judges ix. 13) "in my heart, more than in the time that their corn and (tirosh) wine increased."

We all know that the weary, hungry man is cheered with meat. As soon as the nerves of the stomach are excited by food, a sensation of refreshment, of warmth, and of cheer is felt. The woman who, all day long, has bent over the wash-tub and exhausted her strength, sits down at the close of the day to her cup of tea—

"The cup that cheers, but not inebriates"—

with her frugal meal of bread, and, peradventure, of

meat, and rises up refreshed, cheerful, and strong. We all know that good news is cheering, animating, exhilarating. So, also, is cold water; for thus, saith the Proverb xxv. 25: "As cold water to a thirsty soul, so is good news from a far country." Water, with its cheering power, was the proper illustration.

3. *This wine is the emblem of spiritual blessings.* Isa. lv. 1: " Ho, every one that thirsteth, come ye to the waters, and he that hath no money; come ye, buy, and eat; yea, come, buy wine (yayin) and milk without money and without price." Here the prophet, in the name of God, invites all, every one, to take this wine and milk freely and abundantly. How incongruous to say, Buy milk, and drink abundantly of it, for it is innocent and nutritious, and will do you good; and then to say, Come, buy wine (yayin), an intoxicating beverage, which, if you drink habitually and liberally, will beget the drunkard's appetite, and shut you out of heaven! Can it be that God makes the intoxicating wine the emblem of those spiritual blessings which ensure peace and prosperity in this life, and prepares the recipient for blessedness hereafter? There is harmony between milk and unfermented wine as harmless and nutritious, and they properly stand as the symbols of spiritual mercies. With this view agree the other scriptures cited: Ps. civ. 15: "Wine (yayin) that maketh glad the heart of man;" Judges ix. 13: "Wine (tirosh) which cheereth God and man;" Cant. vii. 9: "Best wine for my beloved;" Prov. ix. 2: "Wisdom hath mingled her wine (yaynah). Come, eat of my bread, and drink of the wine (yayin) I have mingled;" Cant. v. 1: "I have drunk my wine (yayin) with milk: eat, O friends; drink, yea, drink abundantly, O beloved."

Such is the invitation to drink abundantly, because

spiritual blessings never injure, but always do good to the recipient.

4. *This wine is the emblem of the blood of the atonement, by which is the forgiveness of sins and eternal blessedness.* In the institution of the Lord's Supper, as recorded by Matt. xxvi. 26–28 and Mark xiv. 22–24, Christ "took the cup, and gave thanks," saying, "This is my blood of the New Testament," "shed for the remission of sins." The bread and the wine are here united, as in other scriptures, as blessings, but in this case as emblems of the most wonderful manifestation of the divine love to man. Paul, 1 Cor. x. 16: "The cup of blessings which we bless, is it not the communion of the blood of Christ?" At the close, Christ said, "I will not drink henceforth of this fruit of the vine, until that day when I drink it new with you in my Father's kingdom." Thus the cup is associated with the eternal blessedness of the heavenly world. See further comments on Matt. xxvi. 26.

In all the passages where good wine is named, there is no lisp of warning, no intimations of danger, no hint of disapprobation, but always of decided approval.

How bold and strongly marked is the contrast:

The *one* the cause of intoxication, of violence, and of woes.

The *other* the occasion of comfort and of peace.

The *one* the cause of irreligion and of self-destruction.

The *other* the devout offering of piety on the altar of God.

The *one* the symbol of the divine wrath.

The *other* the symbol of spiritual blessings.

The *one* the emblem of eternal damnation.

The *other* the emblem of eternal salvation.

"The distinction in *quality* between the good and the

bad wine is as clear as that between good and bad men, or good and bad wives, or good and bad spirits; for one is the constant subject of warning, designated poison lit-erally, analogically, and figuratively, while the other is commended as refreshing and innocent, which no alcoholic wine is."—*Lees' Appendix*, p. 232.

Can it be that these blessings and curses refer to the same beverage, and that an intoxicating liquor? Does the trumpet give a certain or an uncertain sound? Says Rev. Dr. Nott: "Can the same thing, in the same state, be good and bad; a symbol of wrath, and a symbol of mercy; a thing to be sought after, and a thing to be avoided? Certainly not. And is the Bible, then, inconsistent with itself? No, certainly."—*Nott*, London Ed. p. 48.

Professor M. Stuart, p. 49, says: "My final conclusion is this, viz., that whenever the Scriptures speak of wine as a comfort, a blessing, or a libation to God, and rank it with such articles as corn and oil, they mean, they can mean *only such wine as contained no alcohol that could have a mischievous tendency;* that wherever they denounce it, and connect it with drunkenness and revelling, they can mean only alcoholic or intoxicating wine."

But the position of the advocates of only one kind of wine is that "the juice of the grape, when called wine, was *always* fermented, and, being fermented, was always intoxicating;" "that fermentation is the essence of wine." One exception will destroy the universality of this sweeping statement.

THE WINE OF EGYPT.

Gen. xl. 11: "I took the grapes, and pressed them into Pharaoh's cup, and I gave the cup into Pharaoh's hand."

To break the force of this, it is pleaded that it was only a dream. But a dream designed to certify an immediate coming event could only be intelligible and pertinent by representing an existing usage.

A singular proof of the ancient usage of squeezing the

juice of grapes into a cup has been exhumed at Pompeii." It is that of Bacchus standing by a pedestal, and holding in both hands a large cluster of grapes, and squeezing the juice *into a cup.*

"Plutarch affirms that before the time of Psammetichus, who lived six hundred years before Christ, the Egyptians neither drank fermented wine nor offered it in sacrifice."—*Nott*, Third Lecture.

"In remote antiquity, grapes were brought to the table, and the juice there expressed for immediate use."—*Nott*, London Ed. p. 58.

"Josephus' version of the butler's speech is as follows: He said 'that by the king's permission he pressed the grapes into a goblet, and, having strained the *sweet wine*, he gave it to the king to drink, and that he received it graciously.' Josephus here uses GLEUKOS to designate the expressed juice of the grape before fermentation could possibly commence."—*Bible Commentary*, p. 18.

Bishop Lowth of England, in his *Commentary on Isaiah*, in 1778, remarking upon Isa. v. 2, refers to the case of Pharaoh's butler, and says, "By which it would seem that the Egyptians drank only the fresh juice pressed from the grapes, which was called oinos ampilinos, *i.e.*, wine of the vineyards."

Rev. Dr. Adam Clark, on Gen. xl. 11, says: "From this we find that wine anciently was the mere expressed juice of the grape without fermentation. The saky, or cup-bearer, took the bunch, pressed the juice into the cup, and instantly delivered it into the hands of his master. This was anciently the yayin [wine] of the Hebrews, the oinos [wine] of the Greeks, and the mustum [new fresh wine] of the ancient Latins." Baxter's *Comprehensive Bible* quotes Dr. Clark with approbation.

"It appears that the Mohammedans of Arabia press

the juice of the grape into a cup, and drink it as Pharaoh did."—*Nott*, London Ed. p. 59.

Milton says of Eve :

"For drink the grape she crushed—inoffensive must."

So also Gray :

"Scent the new fragrance of the breathing rose,
And quaff the pendent vintage as it grows."

Nott, 59.

NEW WINE AND OLD BOTTLES.

The first occasion, following the order of the Gospels, on which Christ speaks of wine, he says (Matt. ix. 17): "Neither do men put new wine into old bottles," etc. A similar statement is also made by Mark ii. 22 and Luke v. 37.

Our Lord here refers to a well-known custom, in his day, in relation to the keeping of wine. Notice the facts. They did not put (oinos neos) new wine—the juice fresh from the press—into old bottles, then made of the skins of goats, and the reason is given, "Else the bottles break, and the wine runneth out, and the bottles perish." But it was the custom to put the new wine into new bottles, and the reason is given, "That both the wine and the bottles are thus preserved."

The explanation which the advocates of but one kind of wine give is that new bags were used in order to resist the expansive force of the carbonic acid gas generated by fermentation. This explanation necessarily admits that the new wine had not yet fermented ; for, if it had been fermented, the old bottles would suit just as well as the new ; but the new, it is pleaded, were required to resist the force of fermentation. They thus concede that the new wine had not yet fermented.

Chambers, in his *Cyclopædia*, says: "The force of fermenting wine is very great, being able, if closely stopped up, to burst through the strongest cask." What chance would a goat-skin have?

I have said, if the "new wine" had already fermented, the old bottles would suit just as well as the new; but, if not fermented, the old would not suit, not because they were weak, but because they would have portions of the albuminous matter or yeast adhering to the sides. This, having absorbed oxygen from the air, would become active fermenting matter, and would communicate it to the entire mass.

Liebig informs us that "fermentation depends upon the access of air to the grape-juice, the gluten of which absorbs oxygen and becomes ferment, communicating its own decomposition to the saccharine matter of the grapes."—*Kitto*, ii. 955.

The new bottles or skins, being clean and perfectly free from all ferment, were essential for preserving the fresh unfermented juice, not that their strength might resist the force of fermentation, but, being clean and free from fermenting matter, and closely tied and sealed, so as to exclude the air, the wine would be preserved in the same state in which it was when put into those skins.

Columella, who lived in the days of the Apostles, in his receipt for keeping the wine "*always sweet,*" expressly directs that the newest must, be put in a "*new amphora,*" or jar.

Smith, in his *Greek and Roman Antiquities*, says: "When it was desired to preserve a quantity in the sweet state, an amphora was taken and coated with pitch *within* and *without;* it was filled with the mustum lixivium, and corked, so as to be perfectly air-tight."

The facts stated by Christ are in perfect keeping with the practice prevailing in his day to prevent the pure juice of the grape from fermenting. The *new* amphora—the amphora coated with pitch *within* and *without*—and the *new bottles*, all have reference to the same custom. The people of Palestine must have been familiar with this custom, or Christ would not have used it as an illustration. This passage, properly viewed in connection with the usages of the day, goes a great way toward establishing the fact that Christ and the people of Palestine recognized the existence of two kinds of wine—the fermented and the unfermented.

This passage also helps us to understand the character of the wine Christ used, which he made for the wedding at Cana, and which he selected as the symbol of his atoning blood.

CHRIST EATING AND DRINKING.

Matt. xi. 18, 19 : " John came neither eating nor drinking, and they say, He hath a devil. The Son of man came eating and drinking, and they say, Behold a man gluttonous, and a wine-bibber, a friend of publicans and sinners. But wisdom is justified of her children." The Saviour, in the verses immediately preceding, illustrated the captiousness and unreasonableness of those who were determined not to be pleased, but under all circumstances to find fault. " Whereunto shall I liken this generation ? It is like unto children sitting in the markets and calling unto their fellows, and saying, We have piped unto you, and ye have not danced; we have mourned unto you, and ye have not lamented." Christ directly applies this illustration by reference to the estimate placed upon John and himself by that generation.

John was a Nazarite, and conformed rigidly to the requirements of that order. When they noticed his austere abstinence, peculiar habits, rough attire, and uncompromising denunciations, they were not pleased, and dismissed him with the remark, "He hath a devil." When they saw Christ, whose mission was different from that of John, and perceived that he practised no austerities, but lived like other men, and mingled socially with even the despised of men, they were no better pleased, and said, "Behold a man gluttonous, and a wine-bibber, a friend of publicans and sinners." It is on such authority that the advocates of alcoholic wines claim that Christ was accustomed to use them. At best, it is only inferential, because he ate and drank, and was "a friend of publicans and sinners," that he therefore necessarily drank intoxicating wine. We notice that the same authority which said he was a "wine-bibber" also said he was "gluttonous." And on two other occasions (John i. 20, viii. 48) they said he had a devil. If we believe the first charge on the authority of his enemies, we must also believe the second and the third, for the authority is the same. It will be borne in mind that these, his enemies, traduced his character that they might destroy his influence. They judged that the charge of wine-bibbing, whether it implied drunkenness or sensuality, was the most damaging to his influence as a religious teacher and reformer. It should also be remembered that his enemies were unscrupulous, malignant, and not noted for their truthfulness.

Dr. John J. Owen, in his *Commentary*, says: "As wine was a common beverage in that land of vineyards, in its unfermented state, our Lord most likely drank it." The Saviour did not turn aside from his work to clear himself from the charges which malignity and falsehood brought against him. He simply said, "Wisdom is justified of her children;" that is, My work and my character will ultimately shield me from the power of all false accusations. Those who know me will not be affected by them, and those who hate me will not cease from their calumny.

Matt. xxi. 33 : "Vineyard and wine-press." Neither of these determine anything of the character of the wine which was made. It is begging the question to say that all was fermented, especially as the quotations from ancient authors show that there were two kinds—the fermented and the unfermented.

Matt. xxiv. 38 : "Eating and drinking." These terms denote hilarious, thoughtless, and, perhaps, excessive dissipation. Admit that what they drank was intoxicating, it only proves, what no one denies, that there were inebriating drinks, but does not and cannot prove there were no others.

Matt. xxiv. 49 : "Eat and drink with the drunken." This states a fact which we admit, and is proof that there were then intoxicating liquors, and that some men then used them.

THE LORD'S SUPPER.

Matt. xxvi. 26, 27. Having finished the Passover, our Lord "took bread," unleavened, unfermented bread, and blessed it. This was done always at the Passover, and was by Christ transferred to the Supper. He gave it to his disciples as the symbol of his body. Then he took the cup, and gave thanks. This also was done on giving the third cup at the Passover. This he also transferred, and gave it to his disciples as the symbol of his blood, " shed for the remission of sins." The bread and the cup were used with no discrimination as to their character. To be in harmony with the bread, the cup should also have been unfermented. It was the Passover bread and wine that Christ used. In Ex. xii. 8, 15, 17–20, 34, 39, and other places, all leaven is forbidden at that feast and for seven days. The prohibition against the presence and use of all fermented articles was under the penalty of being " cut off from Israel." " The law forbade *seor*—yeast, ferment, whatever could excite fermentation—and *khahmatz*, whatever had undergone fermentation, or been subject to the action of *seor*."—*Bible Commentary*, p. 280.

Professor Moses Stuart, p. 16, says: " The Hebrew word khahmatz means anything fermented." P. 20: " All leaven, *i.e.* fermentation, was excluded from offerings to God.—Levit. ii. 3–14."

" The great mass of the Jews have ever understood this prohibition as extending to *fermented wine*, or strong drink, as well as to bread. The word is essentially the same which designates the fermentation of bread and that of liquors."

Gesenius, the eminent Hebraist, says that "leaven applied to the wine as really as to the bread."—*Thayer*, p. 71.

The Rev. A. P. Peabody, D.D., in his essay on the Lord's Supper, says: "The writer has satisfied himself, by careful research, that in our Saviour's time the Jews, at least the high ritualists among them, extended the prohibition of leaven *to the principle of fermentation in every form;* and that it was customary, at the Passover festival, for the master of the household to press the contents of 'the cup' *from clusters of grapes preserved for this special purpose."—Monthly Review*, Jan., 1870, p. 41.

"Fermentation is nothing else but the putrefaction of a substance containing no nitrogen. *Ferment*, or *yeast*, is a substance in a state of putrefaction, the atoms of which are in continual motion (*Turner's Chemistry, by Liebig*)." —*Kitto*, ii. 236.

Leaven, because it was corruption, was forbidden as an offering to God. Ex. xxxiv. 25: "Thou shalt not offer the blood of my sacrifice with leaven." But salt, because it prevents corruption and preserves, is required. Levit. ii. 13: "With all thine offerings thou shalt offer salt." If leaven was not allowed with the sacrifices, which were the types of the atoning blood of Christ, how much more would it be a violation of the commandment to allow leaven, or that which was fermented, to be the symbol of the blood of atonement? We cannot imagine that our Lord, in disregard of so positive a command, would admit leaven into the element which was to perpetuate the memory of the sacrifice of himself, of which all the other sacrifices were but types.

Our Lord blessed the bread, and for the cup he gave thanks. Each element alike was the occasion of devout blessing and thanksgiving. This cup contained that which the Saviour, just about to suffer, could bless, and which he, for all time, designated as the symbol of his own atoning blood.

Having finished the Supper, in parting with his disciples he said, "I will not drink henceforth of this fruit of the vine, until that day when I drink it new with you in my Father's kingdom."

The Saviour does not use oinos, the usual word for wine, but adopts the phrase "genneematos tees ampelou," "this fruit of the vine." Was it because oinos was a generic word, including the juice of the grape in all its stages, that he chose a more specific phrase? Was it because he had previously selected the vine as the illustration of himself as the true vine, and his disciples as the fruit-bearing branches, and the juice as "the pure blood of the grape"? (Deut. xxxii. 14.)

By "this fruit of the vine," did he intimate that "in his Father's kingdom" there was something to be looked for there answering to intoxicating wine? This cannot be tolerated for a moment. By "this fruit of the vine," did he mean inebriating wine? Dr. Laurie, *Bibliotheca Sacra*, June, 1869, says, "The Bible never requires the use of wine (intoxicating) except at the communion-table, or as a medicine prescribed by another than the party who is to use it." This is emphatic, and promptly answers the question in the affirmative. It is strange, very strange, that our Lord should require his disciples perpetually to use, as a religious duty, at his table, the article which Dr. Laurie says "all good men agree is dangerous, and not to be used except as a medicine prescribed by another." Does Christ, who has taught us to pray "lead us not into temptation," thus require his disciples to use habitually, in remembrance of him, an article too dangerous to be used anywhere else?

The fact that the Passover was six months later than the vintage is not an invincible objection, since, as we have seen in the preceding pages, on the authority of Josephus, of travellers Niebuhr and Swinburne, and of Peppini, the wine-merchant of Florence, and others, that grapes are preserved fresh through the year, and that wine may be made from them at any period.

Is it probable that Christ took an intoxicating liquor,

which in all the ages past had been the cause of misery and ruin, and which in all the ages to come would destroy myriads in temporal and eternal destruction; that he took the wine which his own inspired Word declared was "the poison of asps," "the poison of serpents," "the poison of dragons," whose deadly bite is like a serpent, and whose fatal sting is like an adder, and made *that* the symbol of his atonement, saying, "*This* is the New Testament in my blood"? But, in "the fruit of the vine," pure, unfermented, healthful, and life-sustaining, and which the Scriptures called "the blood of the grape" and "the pure blood of the grape," there was harmony and force in making it the symbol of atoning blood by which we have spiritual life and eternal blessedness.

The Apostle Paul, 1 Cor. x. 15, not only avoids the word *oinos* (wine), but calls the liquor used "the cup of blessing which we bless, is it not the communion of the blood of Christ?" And in xi. 25 he quotes the exact words of Christ, "This cup is the New Testament in my blood."

Clement, of Alexandria, A.D. 180, designates the liquid used by Christ as "*the blood of the vine.*"—*Kitto*, ii. 801.

Thomas Aquinas says, "Grape-juice has the specific quality of wine, and, therefore, this sacrament may be celebrated with grape-juice."—*Nott*, London Ed. p. 94, note.

Mark ii. 22: "New wine in new bottles." See Matt. ix. 17.

Mark xii. 1: Vineyard, wine-fat. See Matt. xxi. 33.

Mark xiv. 23–25: Lord's Supper. See Matt. xxvi. 26.

Mark xv. 23: "Wine mingled with myrrh." This is a specially prepared article, and not the pure juice of the grape. This Christ refused.

Luke i. 15: "Drink neither wine nor strong drink."

This had reference to John as a Nazarite, and, so far as it is applicable to the case in hand, favors total abstinence as favorable to physical and spiritual strength.

Luke v. 37–39: "New wine in new bottles." See Matt. ix. 17.

Luke vii. 33–35: John the Baptist. See Matt. xi. 18, 19.

Luke x. 7: "Eating and drinking" This direction to his disciples is simply to take of the ordinary hospitality. Only by violent construction can it imply that alcoholic were the only drinks offered them.

Luke x. 34: "Pouring in oil and wine." This was an external and medicinal application. The mixture of the two formed a healing ointment. Pliny mentions "*oleum gleucinum*, which was compounded of oil and gleucus (sweet wine), as an excellent ointment for wounds." "Columella gives the receipt for making it."—*Bible Commentary*, p. 297.

Luke xii. 19: "Eat, drink, and be merry." This is the language of a sensualist, and is used by Christ to illustrate not the propriety of drinking usages, but that covetousness is living to self.

Luke xii. 45: "Eat, drink, and be drunken." See Matt. xxiv. 49.

Luke xvii. 27, 28: "Drank," etc. See Matt. xxiv. 38.

Luke xx. 9: Planted vineyard. See Matt. xxi. 33;

Luke xxi. 34: "Surfeiting and drunkenness," literally, in debauch and drunkenness. Robinson, "properly, *seizure of the head:* hence intoxication."

Christ here warns equally against being "overcharged with surfeiting, and drunkenness, and cares of this life." This text decides nothing in respect to wine which would not intoxicate, but warns against the drinks that would.

Nor does it bear upon the propriety of moderate drinking.

WEDDING-WINE AT CANA.

John ii. 1–11: The distinguishing fact is that Christ turned the water into wine. The Greek word is *oinos;* and it is claimed that therefore the wine was alcoholic and intoxicating. But as *oinos* is a generic word, and, as such, includes all kinds of wine and all stages of the juice of the grape, and sometimes the clusters and even the vine, it is begging the whole question to assert that it was intoxicating. As the narrative is silent on this point, the character of the wine can only be determined by the attendant circumstances—by the occasion, the material used, the person making the wine, and the moral influence of the miracle.

The *occasion* was a wedding convocation. The *material* was water—the same element which the clouds pour down, which the vine draws up from the earth by its roots, and in its passage to the clusters changes into juice. The *operator* was Jesus Christ, the same who, in the beginning, fixed that law by which the vine takes up water and converts it into pure, unfermented juice.

The wine provided by the family was used up, and the mother of Jesus informed him of that fact. He directed that the six water-pots be filled with water. This being done, he commanded to draw and hand it to the master of the feast. He pronounced it wine—good wine.

The moral influence of the miracle will be determined by the character of the wine. It is pertinent to ask, Is it not derogatory to the character of Christ and the teachings of the Bible to suppose that he exerted his miraculous power to produce, according to Alvord, 126, and according to Smith, at least 60 gallons of intoxicating wine?—wine which inspiration had denounced as "a mocker," as "biting like a serpent," and "stinging like an adder," as "the poison of dragons," "the cruel venom of asps," and which the Holy Ghost had selected as the emblem of the wrath of God Almighty? Is it probable that he gave *that* to the guests after they had used the wine provided by the host, and which, it is claimed, was intoxicating?

But wherein was the miracle? We read in Matt. xv. 34 that Christ fed four thousand persons, and in Mark vi. 38 that he fed five thousand persons, in each case upon a few loaves and fishes, taking up seven and twelve baskets of fragments. In these cases, Christ did *instantly* what, by the laws of nature which he had ordained, it would have taken months to grow and ripen into wheat. So in the case of the wine, Christ, by supernatural and superhuman rapidity, produced that marvellous conversion of water into the "pure blood of the grape" which, by his own established law of nature, takes place annually through a series of months, as the vine draws up the water from the earth, and transmutes it into the pure and unfermented juice found in the rich, ripe clusters on the vine.

In Ps. civ. 14, 15, we read: "That he may bring forth food out of the earth, and wine that maketh glad the heart of man." Here the juice of the grape which is produced out of the earth is called wine. This wine was made by the direct law of God—that law by which the vine draws water from the earth and transmutes it into pure juice in the clusters.

I am happy to state that this is not a modern interpretation, forced out by the pressure of the wine question, but was also entertained by the early fathers.

St. Augustine, born A.D. 354, thus explains this miracle: "For he on that marriage-day made wine in the six jars which he ordered to be filled with water—he who now makes it every year in the vines; for, as what the servants had poured into the water-jars was turned into wine by the power of the Lord, so, also, that which the clouds pour forth is turned into wine by the power of the self-same Lord. But we cease to wonder at what is done every year; its very frequency makes astonishment to fail."—*Bible Commentary*, p. 305.

Chrysostom, born A.D. 344, says: "Now, indeed, making plain that it is he who changes into wine the water in the vines and the rain drawn up by the roots. He produced instantly at the wedding-feast that which is formed in the plant during a long course of time."—*Bible Commentary*, p. 305.

Dr. Joseph Hall, Bishop of Norwich, England, in 1600, says: "What doeth he in the ordinary way of nature but turn the watery juice that arises up from the root into wine? He will only do this, now suddenly and at once, which he does usually by sensible degrees."—*Bible Commentary*, p. 305.

The critical Dr. Trench, now Archbishop of Dublin, says: "He who each year prepares the wine in the grape, causing it to drink up and swell with the moisture of earth and heaven, to transmute this into its own nobler juices, concentrated all those slower processes now into the act of a single moment, and accomplished in an instant what ordinarily he does not accomplish but in months."—*Bible Commentary*, p. 305.

We have the highest authority that alcohol is not found

in any living thing, and is not a process of life. Sir Humphry Davy says of alcohol: "It has never been found ready formed in plants."

Count Chaptal, the eminent French chemist, says: "Nature never forms spirituous liquors; she rots the grape upon the branch, but it is *art* which converts the juice into (alcoholic) wine."

Dr. Henry Monroe, in his *Lecture on Medical Jurisprudence*, says: "Alcohol is nowhere to be found in any product of nature; was never created by God; but is essentially an artificial thing prepared by man through the destructive process of fermentation."

Professor Liebig says: "It is contrary to all sober rules of research to regard the vital process of an animal or a plant as the cause of fermentation. The opinion that they take any share in the *morbid* process must be rejected as an hypothesis destitute of all support. In all fungi, analysis has detected the presence of sugar, which during the vital process is not resolved into alcohol and carbonic acid, but AFTER THEIR DEATH. It is the very reverse of the vital process to which this effect must be ascribed. Fermentation, putrefaction, and decay are processes of decomposition." See notes on 1 Tim. iv. 4.

Can it be seriously entertained that Christ should, by his miraculous power, make *alcohol*, an article abundantly proved not to be found in all the ranges of his creation? Can it be believed that he, by making *alcohol*, sanctions the making of it and the giving of it to his creatures, when he, better than all others, knew that it, in the past, had been the cause of the temporal and eternal ruin of myriads, and which, in all the ages to come, would plunge myriads upon myriads into the depths of eternal damnation?

The Rev. Dr. Jacobus says: "All who know of the wines then used, well understand the unfermented juice of the grape. The present wines of Jerusalem and Lebanon, as we tasted them, were commonly boiled and sweet, without intoxicating qualities, such as we here get in liquors called wines. The boiling prevents fermentation. Those were esteemed the best wines which were least strong."—Comments on John ii. 1–11.

This festive occasion furnishes no sanction for the use of the alcoholic wines of commerce at weddings at the present time, much less for the use of them on other occasions.

Acts ii. 13: "Others mocking said, These men are full of new wine."

To account for the strange fact that unlettered Galileans, without previous study, could speak a multitude of languages, the mockers implied they were drunk, and that it was caused by *new wine* (gleukos). Here are two improbabilities. The first is, that drinking alcoholic wine could teach men languages. We know that such wines make men talkative and garrulous; and we also know that their talk is very silly and offensive. In all the ages, and with the intensest desire to discover a royal way to knowledge, no one but these mockers has hit upon alcohol as an immediate and successful teacher of languages.

The second improbability is, that *gleukos*, new wine, would intoxicate. This is the only place in the New Testament where this word occurs. Donnegan's *Lexicon* renders gleukos, "new, unfermented wine—must." From "*glukus*, sweet, agreeable to the taste;" where oinos is understood, "sweet wine made by boiling grapes."

Dr. E. Robinson, quoting classical authorities, defines

gleukos, "must—grape-juice unfermented;" but, seemingly with no other authority than the mockers, adds: "Acts ii. 13: Sweet wine, fermented and intoxicating."

Dr. S. T. Bloomfield says: "Gleukos, not *new-made* wine, which is the *proper* signification of the word (for that is forbidden by the *time of the year*); but *new*, *i.e.* sweet wine, which is very intoxicating."

Rev. T. S. Green's *Lexicon*, *gleukos*, "the unfermented juice of the grape, must; hence, sweet new wine. Acts ii. 13. From glukus, sweet. Jas. iii. 11, 12; Rom. x. 9, 10."

Science teaches that, when by fermentation the sugar is turned into alcohol, the sweetness of the juice is gone. Thus, sweet means, as the lexicons state, unfermented wine.

Kitto, ii. 955, says: "Gleukos, *must*, in common usage, sweet or new wine. It only occurs once in the New Testament (Acts ii. 13). Josephus applies the term to the wine represented as being pressed out of the bunch of grapes by the Archi-oino-choos into the cup of the royal Pharaoh." Professor C. Anthon says: "The sweet, unfermented juice of the grape is termed *gleukos*."

Smith, in his *Greek and Roman Antiquities*, says, "The sweet, unfermented juice of the grape was termed *glūkos* by the Greeks, and *mustum* by the Romans; the latter word being properly an adjective, signifying new or fresh."

Rev. Albert Barnes, on Acts ii. 13, remarks: "New wine (*glukos*)—this word properly means the juice of the grape which distils before a pressure is applied, and called *must*. It was sweet wine, and hence the word in Greek meaning sweet was given to it. The ancients, it is said, had the art of preserving their new wine, with the peculiar flavor before fermentation, for a considerable time, and were in the habit of drinking it in the morning."

Dr. William Smith's *Dictionary of the Bible*, article Wine, says, "A certain amount of juice exuded from the ripe fruit before the treading commenced. This appears to have been kept separate from the rest of the juice, and to have formed the sweet wine (*glukos*, new wine) noticed in Acts ii. 13." "The wine was sometimes preserved in its unfermented state, and drunk as *must*."

It was, indeed, the most consummate irony and effrontery for those mockers to say that the apostles were drunk on *gleukos*, new wine, and full as reliable was the statement that, being thus drunk, they could intelligently and coherently speak in a number of languages of which, up to that day, they had been ignorant. Peter denies the charge, and fortifies his denial by the fact that it was only the third hour of the day, answering to our nine A.M. This was the hour for the morning sacrifice. It was not usual for men to be drunk thus early (1 Thess. v. 7). It was a well-known practice of the Jews not to eat or drink until after the third hour of the day. As distilled spirits were not known until the ninth century, it was altogether an improbable thing that they could have thus early been drunk on the weak wines of Palestine. As the evidence, both ancient and modern, is that *gleukos*, new wine, was unfermented, and therefore not intoxicating, this passage testifies in favor of two kinds of wine.

Acts xxiv. 25, "Reasoned of temperance." "The English word *temperance*," says *Bib. Com.*, p. 317, "is derived directly from the Latin *temperantia*, the root of which is found in the Greek *temō, temnō, tempō,* to cut off. Hence *temperantia* (temperance), as a virtue, is the cutting-off of that which ought not to be retained—self-restraint *from*, not *in*, the use of whatever is pernicious, useless, or dangerous." There is nothing in this text, or its surroundings, which intimates that Paul aimed to persuade Felix to become a *moderate drinker*. The case was more urgent and momentous. This Roman governor of Judea was a licentious man, then living in open adultery; he was an unjust magistrate, and reckless of all retribution except that of Cæsar. Paul, therefore, so probed his conscience with his reasonings upon righteousness, self-control, and responsibility to God, his Creator and final Judge, that he trembled.

Rom. xiii. 13, "Drunkenness." The Greek word *methee* means drunkenness. This was common in Rome, and Paul wisely exhorted the Christians there to avoid it. There is, it will be noticed, the same prohibition of rioting, chambering, and wantonness, as of drunkenness. The argument which uses this text in favor of *moderate drinking* is equally good in favor of *moderate rioting, and chambering, and wantonness, and strife, and envyings*. All agree that in these total abstinence is the only safe and Christian course, and why not equally so in the matter of drunkenness? The best and surest way to avoid drunkenness is to have nothing to do with alcoholic drinks, which produces it, especially as all drunkards are only made out of moderate drinkers.

STUMBLING-BLOCKS.

Rom. xiv. 13, "But judge this rather, that no man put a stumbling-block, or an occasion to fall, in his brother's way." Two words demand examination.

1. *Proskomma*, which Donnegan renders, "Stumble, a trip or false step, an obstacle, an impediment; in general, a hindrance."—*New Testament Lexicon*. Metaphorically, "stumbling-block, an occasion of sinning, means of inducing to sin."—Rom. xiv. 13 and 1 Cor. viii. 9.

2. *Skandalon*. Donnegan, "Cause of offence or scandal."—*New Testament Lexicon*. "Cause or occasion of sinning."

In the context, Paul dissuades from judging one another concerning clean and unclean meats (verses 3 and 14), as a matter of comparatively small moment. But he urges, as a most momentous matter, that Christians should so regulate all their conduct, socially and religiously, as not to put a stumbling-block, or an occasion to fall, in the way of his brother. Thus he establishes a principle of action universally binding in all ages and under all circumstances. This compels every Christian disciple prayerfully to ponder this question, *Do the social drinking usages of the present time put a stumbling-block, or an occasion to fall, in the way?*

No one will maintain, however social they may be, that they are the means of grace, or that they promote spirituality. It must, on the other hand, be admitted that they do circumscribe the usefulness of all, and seriously injure the spirituality of many. No one who uses alcoholic drinks, and furnishes them to his guests, can say they do him no injury. He is not a reliable judge in his own case. Others see and deplore the decline of spirituality and the increased power of worldliness which he makes evident.

The point particularly to be regarded is the influence exerted upon those invited to your festive gatherings, and to whom you offer the intoxicating drinks, even so pressing them as to overcome reluctance, and perhaps conscientious convictions. Do you not thus put a stumbling-block, an impediment, an hindrance, in the way of the Christian usefulness and spiritual progress of your brother —perhaps younger in years, and in the church, than yourself? Do not these prove a cause of offence and of scandal, of sinning and of falling? Where are many who once were active, exemplary members of the churches? Alas! alas! they first learned to sip politely at the fashionable party given by a church member, and by sipping acquired the appetite which led on to drunkenness and the drunkard's grave. We can all recall mournful illustrations.

As others may not have the same cold temperament or self-control as yourself, your example is terrific upon the ardent temperament of the young. For their sakes, the apostolic command binds you to take this stumbling-block, this hindrance, this occasion to sin and to fall, out of the way of your brother. (See Rom. xiv. 17 and xv. 1–3.)

We should never forget what our Lord has said, Matt. xviii. 7, "Woe to the man by whom the offence cometh!" Luke xvii. 1, "But woe unto him through whom they [offences] come! It were better for him that a millstone were hanged about his neck, and he cast into the sea, than that he should offend [cause to stumble or fall] one of these little ones."

I can hardly believe that this subject has been seriously and prayerfully pondered by those Christian professors who habitually spread intoxicating drinks before their guests, especially at evening entertainments, where the young and unsuspecting are convened.

The great barrier which blocks the temperance reform is not found among the drunkards nor in the grog-shops, but in the circles of fashion. So long as these drinks are found in the fashionable parties and defended as the good creatures of God, so long the masses will be so influenced as to be swept along with this fearful tide.

EXPEDIENCY.

Rom. xiv. 14–21, "Neither eat flesh nor drink wine," etc. Expediency necessarily admits the lawfulness and propriety of the use of alcoholic drinks, but that, by reason of the evils which come from the *excessive* use, men should totally abstain. This does not include the idea of personal danger. It rather assumes it as a certainty that the abstainer can so use them as never to exceed the boundaries of prudence. But because of others, not so firm of nerve, or resolute of purpose or power of self-government, we should abstain in order to strengthen, encourage, and save them. In this view, they feel fortified by the noble decision of the Apostle Paul, "Wherefore, if meat make my brother to offend, I will eat no flesh while the world standeth, lest I make my brother to offend." In the Epistle to the Romans, he speaks of those converted from Judaism, but who still felt bound to observe the ceremonial law. Other converts, satisfied that this law was abolished, consequently made no distinction in meats. The former were offended by the practice of the latter. To meet this case, the apostle says, "It is good neither to eat flesh nor drink wine, nor anything whereby thy brother stumbleth, or is offended, or is made weak."

To the Corinthians, 1, viii. 4–13, he speaks of those recently converted from idolatry, and who were troubled about the lawfulness of eating meats which had been offered to idols and then sold in the markets. Whilst he argues that the meat cannot be thus polluted, still, as "there is not in every man that knowledge," and as their weak consciences would be defiled, he admonishes those who were enlightened "to take heed lest by any means this liberty of yours becomes a stumbling-block to them that are weak." He presents the subject in the most solemn and impressive manner, saying, "When ye sin so against the brethren, and wound their weak conscience, ye sin against Christ." The practical and benevolent conclusion to which he comes is, "If meat make my brother to offend, I will eat no flesh while the world standeth, lest I make my brother to offend."

Thus, in two applications, the doctrine of expediency is fully stated. It is necessarily based upon the lawfulness of the usage, and the rightfulness of our liberty in the premises. 1 Cor. x. 23, "All things are lawful unto me, but all things are not expedient: all things are lawful for me, but I will not be brought under the power of any." With Paul, expediency was not the balancing of evils, nor the selfish defence of a doubtful usage; but the law of benevolence, so controlling and circumscribing his liberty as to prevent any injury to the conscience of another. "Even as I please all men in all things, not seeking mine own profit, but the profit of many, that they may be saved."—1 Cor. x. 33.

The abstinence to which Paul alludes was lest the weak conscience of a brother should be wounded. This is not the precise use of the principle in its application to temperance; for those who drink do not plead conscience, and those who abstain do not abstain because for them to drink would wound the consciences of the drinkers. So far from this, our drinking quiets and encourages their consciences. No one can study this argument of the apostle, and his further statement in 1 Cor. ix. 19–23, and fail to feel its benevolent and constraining power. It evolves a principle of action which we are bound to recognize and apply to the necessities of our fellow-men. It demands that we should deny ourselves for the purpose of doing good to others who are exposed to evil. It is the giving up of the use of alcoholic drinks to recover others from ruin, and to save more from taking the first step on the road to drunkenness.

Whilst I fully admit the doctrine of expediency, as laid down by the apostle, 1 am not quite sure that the use which is generally made of it for the cause of temperance may not be turned against us. I am not certain that, as generally expounded, it does not reflect most fearfully, though undesignedly, upon the benevolence of the patriarchs, prophets, the apostles, and even of the blessed Lord our Saviour.

I do not for a moment imagine, much less believe, that the advocates of only alcoholic wines intend to damage the benevolence of the divine Saviour. Yet, when they strenuously claim that he not only personally drank intoxicating wine, but made a large quantity of it for the wedding-guests, they throw shadows over his benevolence; for he, better than all others, knew the seductive and destructive influence of alcoholic drinks, as he could not only look back through all the ages past, but also down through all the ages to come, and tell the myriads upon myriads who by them would be made drunkards and fail of heaven; as he, better than all others, understood the law of benevolence, and knew how to practise self-denial for the good of others. But we hear not one word from him about expediency. What possible claim, then, can this doctrine have upon his followers, if he, with all his wonderfully accurate knowledge, not only did not practise it, but did the reverse, and gave the full force of his personal example for the beverage use of inebriating wines—nay, more, actually employed his divine power in making, for a festive occasion, a large quantity of intoxicating wine? Such is the fearful position in which these alcoholic advocates logically, though unwittingly, place their blessed Lord and ours. But there is no necessity for this dilemma, or for the encouragement it gives to the enemies of temperance. The view we have taken, and, as we trust, proved, satisfactorily explains why neither the patriarchs nor the prophets, why neither Christ nor his apostles, had any occasion to adopt the doctrine of expediency in its application to alcoholic drinks.

The grapes of Palestine being very sweet, and the climate at the vintage season very hot, by the law of fermentation the juice would speedily become sour unless preserved by methods which prevented all fermentation. Having good reason to believe that the wine Christ drank, and which he made for the wedding, was the pure "blood of the grapes," his example gave no sanction to others who used intoxicating wines.

We all are aware that there are many thousands of intelligent Christians who have never yet felt themselves bound by the argument for expediency. They find in it no authority, and it does not bind their conscience. They seize upon the inevitable fact that expediency implies the lawfulness and propriety of the beverage use of alcoholic drinks, and ask, "Why is my liberty judged by another man's conscience?" There are many who seriously doubt whether the reformation can be completed whilst such persons of intelligence and influence are in the way.

At the present time, when there are only alcoholic wines in the walks of commerce, and there is not the choice which, we believe, obtained in the days of Christ, and as these alcoholic beverages are doing wild havoc among men, we fully recognize the law of benevolence as a divine law, and as binding upon every individual. We hold that this law demands that we practise total abstinence, not simply for our own personal safety or that of our family, but especially for the good of others, that they may be rescued from the way of the destroyer, or, what is better, effectually prevented from taking the first step in this road to perdition. We, then, that are strong, ought to bear the infirmities of the weak and not please ourselves. Let every one of us please his neighbor for his good to edification. For even Christ pleased not himself; but, as it is written, "the reproaches of them that reproached thee fell on me."—Rom. xv. 1-3.

1 Cor. vi. 9–11, "Covetous nor drunkards." It will be noticed that drunkards are here classed with fornicators, adulterers, effeminate, thieves, covetous, etc., all of whom, continuing such, "shall not inherit the kingdom of God." Total abstinence *from* all these is a necessity. So long as mere moderation in them is concerned, there is no hope of reformation; nay, so long as any participation in them is concerned, there is no salvation. The moderate use of intoxicating drinks is unsafe; for strong men, in all stations of life, have fallen, and died drunkards, and many are following on. Total abstinence is the scriptural doctrine for all, and from all the practices which expose men to the sins which shut them out of heaven. Christ taught "lead us not into temptation," and Paul exhorts "that no man put a stumbling-block, or an occasion to fall, in his brother's way."

1 Cor. vi. 12, " All things lawful, etc., not expedient." See Rom. xiv. 14–21.

1 Cor. viii. 4–13, "Meat made to offend," etc. See Rom. xiv. 14–21.

1 Cor. x. 22–30, "Sold in shambles, eat," etc. See Rom. xiv. 14–21.

1 Cor. ix. 25, " Temperate." The Greek word *enkratia* is by Donnegan rendered "self-command, self-control, temperance, mastery over the passions;" Robinson and others, N. T. Lexicons, "self-control, continence, temperance." See Acts xxiv. 25, Gal. v. 23, and 2 Pet. i. 6, iv. 5. In the text, it is the power of self-control, or continence, as one striving for the mastery. Dr. Whitby says, " Observing a strict abstinence." Dr. Bloomfield, "extreme temperance and even abstinence." Horace says of the competitor for the Olympic games, " He abstains from Venus and Bacchus." Dr. Clarke states that the regimen included both quantity and quality, carefully abstaining from all things that might render them less able for the combat. The best modern trainers prohibit the use of beer, wine, and spirits. The apostle, having thus illustrated, by reference to the competitors of the Olympic games, his idea of temperance, to wit, *total abstinence*, adds, as an encouragement, " They do it to obtain a corruptible crown, but we an incorruptible." Here is no warrant for moderate drinking, or for those fashionable circles of festivity where the sparkling wines sear the conscience, deaden spirituality, and unfit the Christian professor for that conflict with the world, the flesh, and the devil, the tri-partite alliance which he must overcome, or for ever perish. See Gal. v. 19–23, and notes on Acts xxiv. 25.

1 Cor. xi. 20–34, " Hungry and drunken." " *Methuei*, drunken, being used as antithetical to *peina*, hungry, re-

quires to be understood in the generic sense of *satiated*, and not in the restricted and emphatic sense of intoxicated. That St. Paul should thus have employed it is in harmony with the fact that he was familiar with the LXX. translation of the O. T., where such a use of the word frequently occurs. Gen. xliii. 34, 'Drank and were merry;' Ps. xxiii. 5, 'Cup *runneth over ;*' Ps. xxxvi. 8, 'Abundantly satisfied with the *fatness* of thy house ;' Ps. lxv. 10, 'Settlest the furrows,' *i.e., saturate ;* Jer. xxxi. 14, '*Satiate* the soul of my priests with fatness;' Cant. v. 1, "Drink *abundantly* or be satiated ;' Prov. v. 19, 'Let her breasts *satisfy* thee.' A large collection of such texts, illustraing the usage of *methuō*, will be found in the works of Dr. Lees, vol. ii, showing its application to food, to milk, to water, to blood, to oil, as well as to wine.—*Bib. Com.* p. 340.

Archbishop Newcome, on John ii. 10 and 1 Cor. xi. 21, says, "The word *methuei* does not necessarily denote drunkenness. The word may denote abundance without excess."

Bloomfield, in loco, says, "It is rightly remarked by the ancient commentators that the *ratio oppositi* requires the word to be interpreted only of *satiety* in both drinking and eating. We need not suppose any *drunkenness* or *gluttony*. See *Notes on John* ii. 10. The fault with which they are charged is *sensuality* and *selfishness* at a meal united with the eucharistical feast."—Vol. ii. p. 143.

Donnegan defines *methuō*, "to drink unmixed wine, to drink wine especially at festivals; to be intoxicated; to drink to excess." Robinson, " to be drunk ; to get drunk ; hence, to carouse." Green, " to be intoxicated."

We have thus given a sample of the authorities on the use of this Greek word. It must be plain that the critical students of the New Testament are not all of the opinion that the Corinthian brethren were guilty of drunkenness. Admitting that the word, in this particular place, means to be intoxicated, it proves that there were inebriating drinks, which no one denies, but it cannot prove that these were the *only* kind then used, especially as the word has a generic character and a large application.

The facts of the case are instructive. These converts from idolatry, mistaking the Lord's Supper for a feast, easily fell into their former idolatrous practices. The rich brought plentifully of their viands, and gave themselves selfishly to festivity. The poor, unable thus to provide, were a body by themselves, and were left to go hungry. This discrimination between the rich and the poor was "a despising of the house of God," and was an unchristian act, which the apostle condemned. It is not stated that all the members were drunken, for the narrative expressly says, "One is hungry, and another is drunken," which clearly indicates that a portion were not drunken. As the poor are generally the majority in churches, the strong probability is that a minority only were offenders in prostituting the ordinance and in the matter of drinking. If an intoxicating wine was used on this occasion by the rich church members when they turned the Lord's Supper into a common festive occasion, it furnishes no evidence that such wine was the proper element for the Scriptural celebration of that ordinance. Paul re-enacted the Supper as originally instituted, and restored it to its proper celebration. It is worthy of notice that he says, x. 16, "The cup of blessing which we bless, is it not the communion of the blood of Christ; the bread which we break, is it not the communion of the body of Christ? Ye cannot drink the cup of the Lord and the cup of devils. Ye cannot be partakers of the Lord's table and the table of devils." The contrast between the tables and the cups is apostolic and instructive. Their table and the cup they used were the devil's. The proper table and the proper cup were the Lord's. If their cup contained that which was intoxicating, it was, as Paul declares, the devil's cup; but the cup which contained that which was the opposite, and was not intoxicating, was, as the apostle teaches, the Lord's cup, the cup of blessing.

Gal. v. 19-24, "Drunkenness and temperance." The Apostle Paul draws a striking contrast between the works of the flesh and the fruits of the Spirit. Of the former he says, "Now the works of the flesh are manifest, which are these: adultery, fornication, uncleanness, lasciviousness, idolatry, witchcraft, hatred, variance, emulations, wrath, strife, seditions, heresies, envyings, murders, drunkenness, revellings, and such like; of the which I tell you before, as I have also told you in time past, that they which do such things cannot inherit the kingdom of God."

Of the latter he says, in immediate connection and contrast, "But the fruit of the Spirit is love, joy, peace, longsuffering, gentleness, goodness, faith, meekness, temperance: against such there is no law. And they that are Christ's have crucified the flesh with the affections and lusts." Temperance, which is self-restraint *from*, and not *in*, the use of whatever is injurious, is here placed in opposition to drunkenness. To be safe—abstain. See *Notes on Acts* xxiv. 25 and 1 Cor. ix. 25.

Eph v. 18, "Be not drunk with wine, wherein is excess; but be filled with the Spirit." In this place, *oinos* most probably designates an intoxicating liquor. The word translated excess is *asōtia*, literally *unsavableness*.

It is a word compounded of *a*, privative or negative, and *sōzo*, to save, and thus defined by the lexicon, "The disposition and the life of one who is *asōtos*, abandoned, recklessly debauched, profligacy, dissoluteness, debauchery." Eph. v. 18; Tit. i. 6; 1 Pet. iv. 4.

The apostle here contrasts inebriating wine and the Holy Spirit. He warns men against the wine, and exhorts them to be filled with the Spirit. "He presents a practical antithesis between fulness of wine and fulness of the Divine Spirit; not an antithesis between one state of fulness and another—mere effects—but an antithesis pointing to an *intrinsic contrariety of nature and operation*, between the sources of such fulness, viz., inebriating wine and the Holy Spirit."—*Bib. Com.* p. 353. The excess does not, then, so much apply to the *quantity* of wine used as to the mental and moral condition of the person; since the word *asōtia* denotes such entire dissoluteness of mind and heart as to forbid the hope of salvation.

The apostle properly warns the Ephesian converts against the feasts of Bacchus, where the votaries were made mad by wine and debauching songs; but, in contrast, exhorts them to be filled with the Spirit; and, instead of the noisy, silly talk and songs of the bacchanalians, to manifest their joy and happiness in psalms and hymns and spiritual songs, thus making melody in their hearts unto the Lord.

Olshausen, referring to Luke i. 15, thus comments: "Man feels the want of a strengthening through spiritual influences from without; instead of seeking for these in the Holy Spirit, he, in his blindness, has recourse to the natural spirit, that is, to wine and strong drinks. Therefore, according to the point of view of the Law, the Old Testament recommends abstinence from wine and strong drinks, in order to preserve the soul free from all merely natural influences, and by that means to make it more susceptible of the operations of the Holy Spirit."

The soul filled with the Holy Spirit will not crave an intoxicating beverage to cheer and enliven.

Phil. iv. 5, "Let your moderation be known unto all men. The Lord is at hand." There is not the slightest evidence, either from the original word or the context, that this text has the remotest reference to moderate drinking. The Greek word *epieikees* occurs five times: thrice it is rendered gentle, once patient, and once moderation. In each case, reference is made to the state of the mind, and it might be properly translated, Let your *moderation of mind* be known unto all men. Robinson renders it *meet*, suitable, proper. The reason given for moderation is, "The Lord is at hand." How strange to say to the drinkers, Drink moderately, for the Lord is at hand! But to the Christians at Philippi, then suffering persecutions, the exhortation had point: Let your moderation—that is, your patience, gentleness, mildness, propriety—be known to all men, as a testimony in favor of Christianity. The Lord is at hand is a motive of encouragement.

Col. ii. 16, "Let no man, therefore, judge you in meat or drink," etc.

This has no reference to a distinction of drinks as fermented or unfermented, dangerous or safe, but to those regarded as clean or unclean. That is, proper according to the Jewish law, for the context names holy days, new moon, and Sabbath-days. The point is here—since this law has fulfilled its mission and ceased, therefore use your Christian liberty, and no man must be allowed to condemn you for not now conforming to the requirements of that abrogated law.

1 Thess. v. 7, "They that be drunken are drunken in the night." This simply states a fact in that age, but implies no approbation of intoxicating drinks. The ancient heathen regarded being drunk in the daytime as indecent. In contrast with the stupidity, sensuality, and darkness in which the heathen lived, the exhortation to the Christians who are of the day is to be sober. The Greek word is *nee-phomen*, from *neephoo*, which occurs six times, and is four times rendered sober, and twice watch. The idea of vigilant circumspection and abstinence is impressed by all the context. The classical lexicon defines *nepho* by— "*sobrius sum, vigilo, non bib*,"—to be sober, vigilant, not to drink. Donnegan, "To live abstemiously, to abstain from wine;" metaphorically, "to be sober, discreet, wise, circumspect, or provident, to act with prudence." Robinson's *New Testament Lexicon*, "To be sober, temperate, abstinent, especially in respect of wine." This sobriety is associated with putting on the Christian armor, and it is the call for vigilant wakefulness, having all the powers of mind and body in proper condition.

1 Timothy iii. 2, 3, "Not given to wine." The Apostle Paul, in this first letter to Timothy, whom he calls his "own son in the faith," names thirteen qualifications for a bishop or pastor. "A bishop, then, must be blameless, the husband of one wife, vigilant, sober, of good behavior, given to hospitality, apt to teach; not given to wine, no striker, not greedy of filthy lucre; but patient, not a brawler, not covetous." The language is imperative, "*Must be;*" thus designating that these qualifications are indispensable. He spake with authority, being inspired of God.

It is not my purpose to examine each of these thirteen, but to call attention to three of them, as bearing particularly on the duty of abstinence. In the Authorized Version, we read, "*Vigilant, sober, not given to wine.*" That we may the more perfectly understand the meaning of these, we must look at the original Greek words used by the apostle.

Vigilant.—The Greek is *neephalion,* which Donnegan's *Lexicon* renders "abstemious; that abstains, *especially from wine.*" Hence, "sober, discreet, circumspect, cautious." Robinson's *New Testament Lexicon* defines the word, "Sober, temperate, *especially in respect to wine.*" In N. T., trop., "sober-minded, watchful, circumspect." In the adjective form, the word occurs only in 1 Tim. iii. 2, 11, and Tit. ii. 2, from the verb *neepho,* which Donnegan defines, "To live abstemiously, to abstain from wine." Green's *New Testament Lexicon,* "To be sober, not intoxicated; metaphorically, to be vigilant, circumspect."

Sober.—The Greek is *sōphrona.* Donnegan, "That is, of sound mind and good understanding; sound in intellect, not deranged; intelligent, discreet, prudent, or wise." Green, "Sound; of a sound mind, sane, staid, temperate, discreet, 1 Tim. iii. 2; Tit. i. 8; ii. 3. Modest, chaste, Tit. ii. 5." Macknight, "Sound mind; one who governs his passions, prudent." Bloomfield, "Sober-minded, orderly."

Not given to wine.—The Greek is *mee-paroinon: mee,* a negative particle, *not; paroinon,*compounded of *para,* a preposition governing the genitive (of, from, on the part of), the dative (at, by, near, with), the accusative (together, with, to, towards, by, near, at, next to); and *oinos,* wine. Literally, not at, by, near, or with wine. This looks considerably like *total abstinence.* It applies equally to private habits and public conduct. Notice the careful steps of the progress. He must be *neephalion*, abstinent, sober in body, that he may be *sōphrona*, sound in mind, and that his influence may be unimpaired, *meeparion*, not with or near wine. We find in this passage no countenance for the moderate use of intoxicating wine, but the reverse, the obligation to abstain totally.

"Not given to wine" is certainly a very liberal translation, and shows here the usages of the day unconsciously influenced the translators. "The ancient *paroinos* was a man accustomed to attend drinking-parties." Thus the Christian minister is required not only to be personally sober, but also to withhold his presence and sanction from those assemblies where alcoholic drinks are used, endangering the sobriety of himself and others.

That both Paul and Timothy understood that total abstinence was an essential qualification for the Christian pastor, is evident from the compliance of Timothy. In this same letter, v. 23, Paul advises Timothy, "Drink no longer water, but use a little wine for thy stomach's sake and thine often infirmities." The fact is plain that Timothy, in strict accordance with the direction, "not given to wine," that is, not with or near wine, was a total abstainer. The recommendation to "use a little wine" is exceptional, and strictly medicinal. As there existed in the Roman Empire, in which Timothy travelled, a variety of wines, differing from each other in character, we cannot decide, *ex cathedrâ*, that it was alcoholic wine that Paul recommended. Pliny, Columella, Philo, and others state that many of the wines of their day produced "headaches, dropsy, madness, and *stomach complaints.*" —*Nott*, Lond. Ed. p. 96. We can hardly believe that Paul recommended these. Yet these strikingly designate the effects of alcoholic wines. The same writers tell us that wines destitute of all strength were exceedingly wholesome and useful to the body, *salubre corporis.* Pliny mentions a wine in good repute, *aduminon*—that is, without power, without strength. He particularly states that the wines most adapted to the sick are " *Utilissimum vinum omnibus sacco viribus fractis,*" which the alcoholic wine men translate, " For all the sick, wine is most useful when its forces have been broken by the strainer." We do not object to this rendering, since the wine must be harmless when its forces, which is alcohol, are broken. The Latin word *fractis* is from *frango,* to break in pieces, to dash in pieces, which indicates the thoroughness of the work done by the " sacco," strainer or filter. That the force which the filter breaks is fermentation, is evident from the next sentence of Pliny. (See item " Filtration," on a preceding page.) Horace, lib. i. ode 17, speaks of the *innocentis Lesbii,* innocent Lesbian, which Professor C. Smart renders " unintoxicating." The *Delphin Notes* to Horace say, " The ancients filtered their wines repeatedly before they could have fermented. And thus the *fæces* which nourish the strength of the wine being taken away, they rendered the wine itself more liquid, weaker, lighter, sweeter, and more pleasant to drink."

Again, Horace tells his friend Mæcenas to drink an hundred glasses, without fear of intoxication. (See previous page in this volume.)

Athenæus says of the sweet Lesbian, "Let him take sweet wine (*glukus*), either mixed with water or warmed, especially that called *protropos,* as being very good for the stomach."—*Nott,* Lond. Ed. p. 96, and *Bib. Com.* 374.

Protropos was, according to Pliny, " *Mustum quod sponte profluit antequam uvæ calcentur.*" " The *must* which flows spontaneously from the grapes."—*Nott*, Loud. Ed. p. 80.

Donnegan defines it, "Wine flowing from the grapes before pressure."

Smith's *Greek and Roman Antiquities*, "That which flowed from the clusters, in consequence of their pressure upon each other, to which the inhabitants of Mytelene gave the name *protropos*."

Why not treat Paul with common politeness, not to say honesty, and, as he so emphatically required that a bishop should "not be with or near wine," believe that when he recommended Timothy to "use a little wine" medicinally, he had reference to such wine as Pliny says was "most useful for the sick," whose "forces have been broken by the strainer," or filter? As the recommendation was not for gratification, but for medicine, to Timothy personally, a sick man, and only a little at that, it gives no more countenance for the beverage use of wine for any one, and especially for those in health, than does the prescription of castor-oil by the physician for the beverage use of that article.

The case of Timothy, a total abstainer, illustrates and enforces the inspired declaration that a bishop *must be* vigilant, that is, abstinent; sober, that is, sound in mind; and not given to wine, that is, not with or near wine. If all who are now in the sacred office would follow literally and faithfully the requirements which Paul lays down, "NOT WITH OR NEAR WINE," the number of total abstainers would be greatly increased, the cause of temperance would be essentially promoted, and the good of the community permanently secured; for, according to Paul, total abstinence is an indispensable qualification for a pastor.

1 Tim. iii. 8., Deacons—"not given to much wine."

This is held as evidence not only that they might use some wine, but also that the wine was intoxicating. The Greek word *proseko* occurs twenty-four times, and is eight times rendered beware; six times, take heed; four, gave heed; one, giving heed; two, gave attendance; one, attended; one, had regard; one, given to wine. Robinson's rendering is, "to give or devote one's self to anything;" and other lexicons, "be addicted to, engage in, be occupied with," as in 1 Tim. i. 4; iii. 8. The deacons of the primitive churches were converts mostly from idolatry, and in their unconverted state were accustomed to voluptuousness and sensuality.

In the previous pages, we have seen that those who were dissipated and voluptuous preferred the wine whose strength had been broken by the filter, because it enabled them to drink largely without becoming intoxicated. They used various methods to promote thirst. These voluptuous drinkers continued at times all night at their feasts. "Excessive drinking, even of uninebriating drinks, was a ... e prevalent in the days of St. Paul, and corresponded to gluttony, also common—the excessive use of food, but not of an intoxicating kind."—*Bib. Com.* p. 368. Paul is simply guarding the deacons against a vice of the day.

Such devotion to any kind of wine showed a voluptuousness unseemly in one holding office in the church of Christ. "To argue that, forbidding much wine, Paul approves of the use of some wine, and of any and every sort, is to adopt a mode of interpretation dangerous and wholly inconsistent with common usage." When applied to the clause, "'not greedy of filthy lucre,' it would sanction all avarice and trade craftiness short of that greed which is mean and reckless." But Paul, and other inspired writers, make all covetousness to be idolatry, and not to be once named, much less practised by the saints, even moderately.

1 Tim. iii. 11, " Wives, be sober." The same Greek word is in verse 2 rendered vigilant, and which Donnegan renders abstemious, that abstains, especially from wine. The N. T. Greek lexicons define it, " temperate, abstinent in respect to wine."

1 Tim. iv. 4, "Every creature of God is good," etc. This text has no reference to drinks of any kind, but is directly connected with the meats named in verse 3, and which some had forbidden to be eaten. These, the apostle says, are to be received and used, because they are the creatures of God, and by him given for the good of man. The original word *broma* occurs seventeen times, and is always rendered meat and meats, except once, victuals. Robinson, *eatables, food, i.e.,* solid food opposed to milk. 1 Cor. iii. 2. It means food of all kinds proper to be eaten. But alcohol is not meat in any sense. It is not food; it will not assimilate, nor does it incorporate itself with any part of the body. Says Dr. Lionel S. Beale, Physician to King's College Hospital, England, "Alcohol does not act as food ; it does not nourish tissues." Dr. James Edmunds, of Edinburgh, says, "Alcohol is, in fact, treated by the human system *not as food, but as an intruder and as a poison.*"

In keeping with this is the statement, 1 Sam. xxv. 37, "When the wine was gone out of Nabul." This is singularly accurate, and accords with the most approved discoveries of science, viz., " that intoxication passes off because the alcohol goes out of the body—being expelled from it by all the excretory organs as an intruder into and disturber of the living house which God has fearfully and wonderfully made." Dr. Willard Parker, of New York, has used the same illustration.

The testimony of Dr. J. W. Beaumont, Lecturer on *Materia Mediça* in Sheffield Medical School, England, is, "Alcoholic liquors are not nutritious, they are not a tonic, they are not beneficial in any sense of the word."

The original grant for food reads, Gen. i. 29, " Behold, I have given you every herb bearing seed which is upon the face of the earth, and every tree in the which is the fruit of a tree yielding seed; to you it shall be for meat." Verse 31 : "And God saw everything that he had made, and behold it was very good."

The original grant extended only to vegetables. These were for meat, literally "for *eating*," or that which is to be eaten. Every direct product of the earth fit for food is here given to man. The design was to sustain life. Hence, whatever will not assimilate and repair the waste is not food, and not proper for the use of man.

Who imagines, when the work of creation was finished, that alcohol could then be found in any living thing fresh from the hand of the Creator? God, by his direct act, does not make alcohol. The laws of nature, if left to themselves, do not produce it. By these laws, the grapes ripen; if not eaten, they rot and are decomposed. The manufacture of alcohol is wholly man's device. The assertion that alcohol is in sugar, and in all unfermented saccharine substances which are nutritious, is contradicted by chemical science. The saccharine matter is nutritious, but fermentation changes the sugar into alcohol, by which process all the sugar is destroyed, and, as the alcohol contains no nitrogen, it cannot make blood or help to repair bodily waste. The testimony of eminent chemists is very decided. Sir Humphry Davy, in his *Agricultural Chemistry*, says of alcohol, "It has never been found ready formed in plants." Count Chaptal, the great French chemist, says, "Nature never forms spirituous liquors; she rots the grape upon the branches, but it is art which converts the juice into (alcoholic) wine."

Professor Turner, in his *Chemistry*, affirms the non-natural character of alcohol, "It does *not* exist *ready formed in plants*, but is a product of the vinous fermentation—a process which must be initiated, superintended, and, at a certain state, arrested by art."—*Bib. Com.* p. 370.

Dr. Henry Morrison, of England, in his *Lecture on Medical Jurisprudence*, says, "Alcohol is nowhere to be found in any product of nature, was never created by God, but is essentially an artificial thing prepared by man through the destructive process of fermentation." *

---

* The four following experiments tell their own tale:

"1. One pound of fully ripe grapes (black Hamburgs) were put into a glass retort, with half a pint of water, and distilled very slowly, until three fluid-ounces had passed into the receiver. This product had no alcoholic smell. It was put into a small glass retort, with an ounce of fused chloride of calcium, and distilled very slowly, till a quarter fluid-ounce was drawn; this second educt had no smell of alcohol; nor was it, in the slightest degree, inflammable."

"2 and 3. A flask was filled with grapes, none of which had been deprived of their stalks, and it was inverted in mercury. Another flask was filled with grapes from which the stalks had been pulled, and many of them otherwise were bruised. This flask was also inverted in mercury. The flasks were placed, for five days, in a room of the average temperature of about 70°.

"In the perfect grapes no change was perceivable. In the bruised grapes, *putrefaction* had proceeded to an extent, in each grape, pro-

For the views of Professor Liebig on fermentation, see "Fermentation," in this treatise.

"Lo, this have I found," saith the wise man (Ecc. vii. 29), "that God made man upright, but they have sought out many inventions."

The things created for food, and which are to be received with thanksgiving, are those which are in their natural and wholesome condition, and which nourish and strengthen the body, and not those which are in the process of decomposition. Rotten fruits of all kinds are rejected as innutritious and unwholesome. So also are decaying meats. It is a strange perversion of all science, as well as of common sense, to rank among the good creatures of God alcohol, which is found in no living plant, but which is to be found only after the death of the fruit, and is the product of decomposition.

portionate to the degree of injury it had sustained; the sound parts of each continued unchanged."

"4. The grapes were now removed from the flasks, and the juice expressed from each. The juice from the bruised grapes had not an alcoholic, but a *putrescent* flavor. The juice from the sound grapes was perfectly sweet.

"Both these juices were placed in tightly corked phials half-filled, and subjected to a proper fermenting temperature. IT WAS THREE DAYS before the COMMENCEMENT of fermentation, in each, was indicated by the evolution of carbonic acid gas, as also by the color of the alcohol, and of the aromatic oils always generated in such cases. I, therefore, still believe it to be a FACT that grapes do not produce alcohol; that it can result only where the juice has been expressed from them, and then not *suddenly ;* and that, where the hand of man interferes not, alcohol is never formed."—S. Spence, Chemist to the Yorkshire Agricultural Society; F. R. Lees, Appendix B, pp. 198 and 199.

These justify the statement of Mr. Lees, that "neither ripened nor rotting grapes ever contain alcohol."

The analysis of wines, as published in the *Lancet*, Oct. 26, 1867, shows that, in one thousand grains of the wines named, there was only one and one-half grains of albuminous matter, whilst in the same amount of raw beef there were two hundred and seven grains, that is, one hundred and fifty-six times more nourishment in the same quantity of beef than in wine.—*Bib. Com.* p. 370. The analysis of the beer in common use proves that there is more nourishment in one small loaf of wheat bread than in many gallons of beer. Medical men testify that the flesh of habitual beer-drinkers becomes so poisoned that slight wounds become incurable, and result often in speedy death.

1 Tim. v. 23, "No longer water." See 1 Tim. iii. 2, 3. Titus i. 7, 8, "Not given to wine," "temperate."

Here Paul mentions the same qualifications for a pastor as those stated in his first letter to Timothy iii. 3, "Not given to wine." He uses the same Greek word, *meeparoinon*, compounded of *mee*, a negative particle, *para*, a preposition, with or near, and *oinon*, wine, meaning not near wine, which is a happy apostolic definition of total abstinence. He adds temperate, which, it is pleaded, sanctions moderate drinking. The Greek word here used is *enkratees*. Donnegan, "Holding firm, mastering one's appetite or passions."—*New Testament Lexicon*. "Strong, stout, possessed of mastery, master of self."—Tit. i. 8. It is clear that Paul does not contradict himself in this verse: first, by saying the bishop must be a total abstainer—*mee*, not; *para*, near; *oinon*, wine—and then, in the second place, by saying he must be a moderate drinker. What he here means by temperance applies to the mind and not to the bodily habits. Or if it is contended that it does refer to the body, then it means what he says in

1 Cor. ix. 25, where he uses the same word in reference to those contending for the mastery in the games. Such abstain totally from wine and all excitements, or as Horace expresses it, "He abstains from Venus and Bacchus." See Note, 1 Cor. ix. 25 and Acts xxiv. 25.

Titus ii. 2, 3. The aged men are exhorted to be *sober*, "temperate." The Greek is *ncephalion*, "sober, temperate, abstinent in respect to wine." In N. T., metaphorically, "vigilant, circumspect."—1 Tim. iii. 2, 11; Tit. ii. 2. For temperate the Greek is *sophronos*, "sound of mind, sober-minded, sedate, staid." Temperate, see note on Tit. i. 8.

In verse 3 the aged women are exhorted, "not given to much wine." See comment on 1 Tim. iii. 8.

These were to teach the young women to be "sober." Here the same original word is used which denotes sober-mindedness. See comment on 1 Tim. iii. 2. The necessity of such an exhortation is obvious from the fact that these, before their conversion, had been idolaters, and who, in the days of their ignorance, had given themselves up to voluptuous practices.

Polybius, in a fragment of his 6th book, says, "Among the Romans, the women were forbidden to drink (intoxicating) wine; they drink, however, what is called *passum*, made from raisins, which drink very much resembles *Ægosthenian and Cretan gleukos* (sweet wine), which men use for allaying excessive thirst."—*Nott*, London Ed. p. 80. See notes John ii. 1–11.

Westein commenting on Acts ii. 13, *glukus*, new sweet wine, says, "The Roman ladies were so fond of it, that they would first fill their stomachs with it, then throw it off by emetics, and repeat the draught.—*Bib. Com.* p. 378.

Dr. F. R. Lees says, in the same page, "We have referred to Lucian for ourselves, and find the following illustration: 'I came, by Jove, as those who drink *gleukos* (sweet wine), swelling out their stomach, require an emetic.'" These voluptuous habits denoted such a devotion to the enjoyment of luxury and pleasure, such an indulgence in sensual gratification, as unfitted these women for a station in the Christian church, and for the proper discharge of the domestic duties particularly noticed in the text.

The Rev. W. H. Rule, in his brief enquiry, speaking of this unfermented wine, says: " A larger quantity might be taken, and the eastern sot could enjoy himself longer over the cup, than if he were filled up with fermented wine, without being baffled by the senselessness of profound inebriation.—*Nott*, Lond. Ed. p. 223. Mr. Rule, though no particular friend to the temperance cause, here concedes the fact that there were two kinds of wine, the fermented and the unfermented.

1 Peter i. 13, "Be *sober*." See comments on 1 Thess. v. 7, p. 25, where the same word occurs.

1 Peter iv. 1–5, "Excess of wine, excess of riot." In this passage three facts are significant and instructive. The first is, that in their unconverted state these converts whom Peter addresses lived in the lusts of men, wrought the will of the Gentiles, and walked in lasciviousness, lusts, excess of wine, revellings, banquetings, and abominable idolatries. The second fact is, that their former companions thought it strange that, being Christians, they would not "run with them to the same excess of riot." The third fact is, that their former companions spoke evil of them because of their abstinence.

It is clear that the Christians named in this passage

were abstainers from their former usages, and that on this account they were spoken evil of, very much as are the total abstainers of the present day.

Oinophlugia occurs only in this text, and is a compound of *oinos*, wine, and *phluō*, to overflow = a debauch with wine. Probably intoxicating, though the wine broken by the filter was preferred by the voluptuous and dissipated.

The Greek word *asōtia*, in Eph. v. 18, is rendered excess, and is connected with wine; and means, literally, unsavableness, utter depravity, and dissoluteness. In the text, and Tit. i. 6, it is connected with riot, which means overflow, outpouring of dissoluteness, thus denoting the same moral character. As the two phrases occur in the text, it teaches that excess of wine and excess of riot are related to each other as cause and effect; but excess of wine no more justifies moderate drinking than excess of riot justifies moderate rioting. The design of Peter was to encourage those to whom he wrote to continue in their abstinence.

1 Peter iv. 7, "Be ye therefore *sober*." See 1 Tim. iii. 2. The motive for sober-mindedness is the same as Phil. iv. 5, which see.

1 Peter v. 8, "Sober, vigilant. See 1 Thess. v. 6–8; Tit. ii. 2; 1 Peter i. 13 and iv. 7. The sobriety here has no reference to intoxication, but to the state of mind according with vigilance. The reason for wakeful vigilance is the activity and malignity of the devil.

2 Peter i. 6, "Temperance." See Acts xxiv. 25; 1 Cor. ix. 25; and Gal. v. 25.

In the Revelation there are nine references to wine. In chap. vi. 6 and xviii. 13, wine and oil are mentioned as articles of necessary comfort and merchandise. In xiv.

8, xvii. 2, and xviii. 3, we read, "Wine of the wrath of her fornication," "Drunk with the wine of her fornication," and "Drunk of the wine of her fornication." These are figurative, and imply punishment. In xiv. 10, "Drink of the wine of the wrath of God;" xvi. 19, "Cup of the wine of the fierceness of his wrath." In xiv. 19, "Great wine-press of the wrath of God," and xix. 15, "Wine-press of the fierceness and wrath of Almighty God." These are symbols of the divine vengeance.

I have now called attention to every passage in the New Testament where wine is mentioned, and have given to each that interpretation which to me appeared just and proper. How far I have carried the full conviction of my readers, each one must determine for himself. The results recorded in these pages have cost me years of patient and laborious investigations. My own convictions have steadily deepened and become firmer as I have canvassed the positions maintained by writers who hold views widely differing from my own. This, some may think, is stubborn obstinacy on my part; but I do not thus judge myself, as I am conscious, however I may err, of desiring only to know the truth, and hold such an understanding of the Bible as will best harmonize the law of God as developed by true science, and the law of God as written in the inspired page.

I do not say that there are no difficulties connected with the wine question. All I ask is that the students of the Bible will treat these with the same candor and desire to harmonize them that they do the difficulties connected with astronomy, geology, and conflicting historical statements. If the language of the Bible can be honestly so interpreted as to harmonize with the undisputed facts developed by the temperance reformation, in relation to the effects of alcoholic drinks, with the testimony of the most intelligent physicians and eminent chemists, that alcohol contains no nourishment, will neither make blood nor repair the waste of the body, but it is an intruder and a poison; this will secure the firm friendship of many who now stand aloof, and will promote the temporal, spiritual, and eternal happiness of mankind.

TESTIMONY.

The following testimony, from four eminent scholars, may fortify the convictions already produced by the facts and reasonings found upon the preceding pages:

PROFESSOR GEORGE BUSH.—Mr. E. C. Delavan, having been referred to Professor Bush, as a learned Biblical scholar, from whom he might obtain correct information as to Bible temperance, visited him in his library, and stated to him his views on the wine question. With promptness he condemned them, and, referring to a text, he said, "This verse upsets your theory." When asked to refer to the *original,* he did so, and, with amazement, said, "*No permission to drink intoxicating wine here. I do not care about wine, and it is very seldom that I taste it, but I have felt until now at liberty to drink, in moderation, from this verse.*" Being entreated to make this a subject of special and particular examination, he said he would. At a subsequent visit he thus greeted Mr. Delavan: "*You have the whole ground, and, in time, the whole Christian world will be obliged to adopt your views.*" At the request of Mr. Delavan, he published his views in the *New York Observer* (*Enquirer,* Aug., 1869). This testimony is the more valuable, as it is not only the result of a careful examination of the original languages, but the honest surrender to the force of evidence of a previous conviction.

Rev. Dr. E. Nott, late President of Union College, in his fourth lecture says: "That unintoxicating wines existed from remote antiquity, and were held in high estimation by the wise and good, there can be no reasonable doubt. The evidence is unequivocal and plenary." "We know that then, as now, inebriety existed; and then, as now, the taste for inebriating wines may have been the prevalent taste, and intoxicating wines the popular wines. Still unintoxicating wines existed, and there were men who preferred such wines, and who have left on record the avowal of that preference."—*Nott*, Lon. Ed. p. 85.

Professor Moses Stuart.—" My final conclusion is this, viz.: that, whenever the Scriptures speak of wine as a comfort, a blessing, or a libation to God, and rank it with such articles as corn and oil, they mean—they can mean—*only such wine as contained no alcohol that could have a mischievous tendency;* that wherein they denounce it, prohibit it, and connect it with drunkenness and revelling, they can mean *only alcoholic* or *intoxicating wine*.

"If I take the position that God's *Word* and *works* entirely harmonize, I must take the position that the case before us is as I have represented it to be. Facts show that the ancients not only preserved wine unfermented, but regarded it as of a higher flavor and finer quality than fermented wine. Facts show that it was, and might be, drunk at pleasure without any inebriation whatever. On the other hand, facts show that any considerable quantity of fermented wine did and would produce inebriation; and also that a tendency towards it, or a disturbance of the fine tissues of the physical system, was and would be produced by even a small quantity of it; full surely if this was often drunk.

"What, then, is the difficulty in taking the position that the *good and innocent wine* is meant in all cases where it is commended and allowed; or that the *alcoholic* or *intoxicating wine* is meant in all cases of prohibition and denunciation?

"I cannot refuse to take this position without virtually impeaching the Scriptures of contradiction or inconsistency. I cannot admit that God has given liberty to persons in health to drink alcoholic wine, without admitting that his *Word* and his *works* are at variance. The law against such drinking, which he has enstamped on our nature, stands out prominently—read and assented to by all sober and thinking men—is his Word now at variance with this? Without reserve, I am prepared to answer in the negative."

It was after an exhaustive examination, the details of which are contained in his printed letter of sixty-four pages octavo, that he gave to the world this full and unequivocal testimony we have just recited.

REV. ALBERT BARNES, in his commentary on John ii. 10, says: "The wine of Judea was the pure juice of the grape, without any mixture of alcohol, and commonly weak and harmless. It was the common drink of the people, and did not tend to produce intoxication."

All acquainted with Mr. Barnes know that he would not make such a statement until he had given the subject a patient and thorough examination. Having scrutinized all the authorities, he has thus recorded upon the printed page his clear and honest convictions.

Beside these testimonies, a goodly number of men, well

read in ancient lore and learned in the original languages of the Word of God, have, by patient study, been led to the same conclusion. The company of such is rapidly increasing both in Great Britain and America. We do not despair, but confidently believe that the time is not far distant when no drinker, nor vender, nor defender of alcoholic wines, will find a shelter and a house of refuge in the Scriptures of God. LET THERE BE LIGHT!

INDEX.

Acetous, 16, 18, 25.
Adams's Antiquities, 29, 38, 39.
Æneus, 41.
Ægosthenian, 44.
Aged men, 117.
Ahsis, 59.
Aigleuces, 37.
Aintab, 32.
Albanian, 41.
Alcohol, 15, 17, 88, 114 note.
Aleppo, 30.
Allar, 66.
Alsop, Robt., 30.
Alvord, 86.
Amethyston wine, 42.
Amphora, 37, 43, 77.
Authon's Dict., 15, 30, 34, 35, 38, 46, 47, 90.
Antiquities, Adams's, 29, 38, 39.
" Smith's, 29, 34, 37, 45, 46, 76, 90, 110.
Aquinas, Thomas, 83.
Arabia, 61.
Aristotle, 27, 42, 45.
Arcadia wine, 27.
Ashishah, 60.
Asia Minor, 18, 32.
Athenæus, 43, 49, 109.
Attic honey, 49.
August, 21.
Augustine, 87.
Authorized Version, 53.

Bacchus, figure of, 73.
Bacchus, 12, 29, 73.
" Anti-, 12, 16, 18, 25, 37, 41.
Baptist, John, 77.
Bad wine, 63.
Barnes, Albert, 63, 90, 123.
Barry, E., 40, 50.
Beale, Dr., 112.
Beaumont, Dr., 113.
Blessings, 67, 68, 70.
Bible and Science, 13, 52.
" Commentary, 9.
Bibline sweet, 43.
Bishops, 106.
Bloomfield, S. T., 90, 100, 101.
Blood, emblem, 71.
" of vine, 80.
Boerhave, H., 26.
Boiled wine, 24, 26, 32, 59.
Bottles, 64, 75.
Bowring, Dr., 29.

Brown, W. G., 28.
Bush, Prof., 121.
Calmet, A., 25.
Calabrians, 35.
Cana wedding, 85.
Candia, 30.
Carenum, 29.
Carr, T. S., 38, 40.
Castratum, 33.
Cato, 48.
Chaldee, 61.
Chambers's Cyclopædia, 14, 76.
Chaptal, Count, 17, 88.
Chemical science, 22, 26, 36, 39, 51.
Chrysostom, 87.
Cider, 41.
Clark, Adam, 74.
" D. E., 21.
Classification of texts, 12, 62.
Clement, 83.
Climate, 18, 19.
Columella, 27, 37, 38, 42, 48, 76.
Concessions, 11.
Contrast of texts, 71, 72.
Cook, Capt., 19.
Corlaer's Hook, 10.
Corn, 67, 69.
Covetous, 99.
Creature of God, 112.
Crete, 30.

Damascus, 32.
Dandalo, Count, 40.
Dandini, 21.
Dates, 55.
Davy, Sir Humphry, 88, 114.
Deacons, 111.
Decomposition, 18.
Defrutum, 27, 30, 59.
Delavan, E. C., 28, 121.
Delphin Notes, 35, 43, 109.
Democritus, 27.
Depurating, 38.
Dibbs, 30, 31.
Diehl, 19.
Distillation, 33, 55.
Donavan, 16, 24, 27, 33.
Donnegan, 89, 93, 101, 106, 107, 110, 116.
Drinks, 105.
Drinking, Christ, 77.
Drugs, 47.
Drunkards, 99, 108.
Drunken, 92, 100, 106.
Duff, Dr. H., 24.

Eating, Christ, 77.
Edmunds, Dr., 112.
Effeminatum, 33.
Elsworth, Hon. O., 44.
Encyclopædia, London, 40.
Engedi, 20, 22.
English generic words, 60.
Epsuma, 29.
Eshcol, 20, 22.
Excess, 118.
Expediency, 95.
Experiments, 114 note.
Exploration, 10.
Eunuchrum, 33.

Fabroni, 17, 19.
Fermentation, 15, 30.
" prevented, 24.
First-fruits, 66.
" sermon, 10.
Filtration, 24, 33–36.
Florence, 23.
Fruit cake, 60.
" preserved, 23.
" sweet, 18.
" of vine. 82.
Fumigation, 39, 43.

Gall, 64.
Gardiner, 39.
Generic words, 54, 60.
Gesenius, 80.
Gill, Dr., 64.
Glenkos, 14, 15, 44, 46, 74, 89, 90, 91, 117.
Glukus, 42, 43, 89, 117.
Gleuxis, 29.
Gluten, 16, 17, 24, 33, 34, 36, 37, 57.
Good wine, 66.
Grapes, 18.
" juice, 24, 26, 27, 31–34, 46, 51, 54, 59, 73, 74, 83.
" molasses, 31.
Gray, 75.
Greek words, 60.
Green, T. S., 90, 101.

Hall, Joseph, 87.
Harmer, 39.
Hebrew words, other, 59.
Helbon, 30.
Helen, 47.
Henderson, 44.
Henry, Matthew, 45.
Hepsinia, 59.
Herod, 23.
Hesiod, 49.
Hippocrates, 43, 49.
History, 9.
Holmes, Rev. Henry, 31.
Homer, 42, 43, 47, 49.
Honey, Attic. 49.
Horace, 28, 35, 42, 100, 109.
Horn, Hartwell, 21.
Hot climate, 19.

Inspired original text, 8, 54.
Inspissated, 26.
Intoxicating wine, 31, 32, 35, 61.

Introduction, 7.
Italy, 35, 49.

Jacobus, Dr., 32, 89.
Jahn. Dr., 20.
Jericho, 21.
John Baptist, 77.
Johnson's Dictionary, 15.
Josephus, 23, 74, 90.
Judging, 105.
Juice, 73, 74.
June, 21.

Kesroan, 30.
Khamah, 64.
Khahmatz, 80.
Khemer, 58.
Kitto, 18, 19, 29, 90, 55, 56, 57, 90.
Koht, J. G., 48.

Lacedæmonians, 27.
Lancet, analysis of wine, 116.
Latin words, 60.
Laurie, Dr., 46, 61, 82.
Leaven, 80, 81.
Lebanon, 21, 29, 30.
Lees, Dr., 9, 36, 51, 54, 56, 58, 59, 61, 72, 118.
Leiber, 30.
Lewis, Taylor, 9, 54.
Lightfoot, 5!.
Liebig, 17, 25, 26, 59, 76, 81, 85, 115.
Littleton's Dictionary, 14.
Lixivium. 37.
Lord's Supper, 79.
Lowth, Bishop, 74.

Mandelslo, 18.
Mamlaqqim, 60.
Mariti, 21.
Masada, 23.
Meats, 105.
Medical enquiries, 45.
Meronian, 49.
Mesek, 59.
Mill, J. S., 62.
Milton, 75.
Miracle, 86.
Mixed wine, 50.
Mocking, 89.
Moderation, 105.
Monroe, Henry, 88.
Morrison, Dr., 114.
Mullen, Dr., 19.
Murphy, Dr., 55.
Must, 27, 37, 38, 39, 41, 45, 91.
Mytelene, 45.

Natural taste, 22.
Nau, 21.
Nazarite, 77, 78.
Neitchutz, 21.
Nenman, C., 29.
Nevin, J. W., 22.
New bottles, 75.
" wine, 75 89.
Newcome, Archbishop, 64, 101.
Nicander, 41.
Nicochaus. 49.
Niebuhr, 82.

INDEX.   127

Nitrogen, 81.
Nordheimer, Prof., 64.
Nott's Lectures, 8, 54, 72, 74, 122.

October, 21.
Offerings, 66.
Oil, 66.
Oinos, 14, 41, 43, 61, 74, 81.
Old bottles, 75.
" wine, 27.
Oleum gleucinum, 38.
Olshausen, 104.
Olympic games, 100.
Opimian wine, 27.
Originality, 7.
Original text inspired, 8, 54.
Owen, D. J., 78.
Palladius, 27.
Palestine, 19, 20, 22, 23.
Palm, 18, 56.
Passover, 51, 79.
Passum, 44.
Parker, Dr. W., 113.
Parkhurst, 64.
Parkinson, 26.
Peabody, A. P., 80.
Pereira, Dr., 17.
Pharaoh, 72.
Pippini, Senior, 23.
Plato, 45.
Plautus, 41.
Pliny, 23, 34, 35, 36, 41, 48, 59, 108, 109, 110.
Plutarch, 34, 45, 74.
Poison, 63, 64.
Polybius, 44, 117.
Portland, Duke of, 21.
Potter, Archbishop, 27.
Pramnian wine, 49.
Preserving fruits, 23.
"   wine, 25.
Prohibitory Roman law, 45.
Protopos, 44, 45, 109, 110.

Question, the, 13.

Reading Cyrus, 30, 40.
Receipts for wine, 37, 38, 39, 47.
Rees' Cyclopædia, 14.
Retimo, 30.
Robinson, E., 84, 89, 100, 106, 112.
Robson, Smylie, 32.
Rockingham, Marquis, 21.
Roman prohibitory law, 45.
"   wines, 27, 39, 41, 46.
"   women, 44.
Rule, W. H., 46, 118.
Russell, Dr. A., 30.

Sabc, 59.
Salt, 81.
Sapa, 27, 30, 59.
Science, 13, 22, 26, 36, 90.
Scriptures, 53.

Gen. xxvii. 28, 37,		67
xxix. 11,		20
xl. 11,		74
xlix. 11,		19
Exodus xii. 8, 39,		80

Scriptures:

Exodus xii. 42,		60
xxxiv. 25,		81
Levit. ii. 11,		67
x. 9,		60
Numb. vi. 3,		60
xiii. 24,		20
xviii. 12,		66
xxviii. 7,		60
Deut. vii. 13,		67
viii. 7,		20
xi. 14,		68
xiv. 26,		60
xxix. 6,		60
xxxii. 14,	20,	59
xxxii. 24,		64
Judges ix. 13,	69,	70
xiii. 4, 7, 14,		60
1 Sam. i. 15,		60
2 Sam. vi. 10,		60
2 Kings xviii. 32,		20
1 Chron. xvi. 3,		60
Ezra vi. 9,		59
vii. 22,		59
Nehem. viii. 10,		60
x. 37,		66
x. 39,	57,	67
xiii. 5, 12,		57
Job vi. 4,		64
Psalms iv. 7,		69
lviii. 4,		64
lx. 3,		65
lxix. 12,		60
lxxv. 8,	48, 50, 59, 60,	65
cii. 9,		60
civ. 14	69, 70,	86
cxl. 3,		64
Prov. iii. 10,	57,	68
iv. 17,		63
ix. 2–5,	50, 60,	70
xx. 1,		60
xxiii. 29–31,	59,	63
xxv. 25,		70
Cant. ii. 5,		60
v. 1,		70
vii. 9,		70
Isaiah i. 22,		59
v. 2,		74
v. 11,		60
v. 22,	47, 60,	65
xxiv. 7,		57
xxiv. 9,		60
xxvii. 2,	20,	59
xxviii. 7,		60
xxix. 9,		60
li. 17,		65
lv. 1,		70
lvi. 12,	60,	63
lxii. 8,	57,	68
lxv. 8,	57,	68
Jeremiah xxv. 15,	48,	65
xlviii. 11,		60
Dan. v. 1–4,		59
Hosea iii. 1,		60
iv. 14,		57
iv. 18,		59
vii. 5,		64
ix. 2,		57

INDEX.

Scriptures:
- Joel i. 5, — 59
- i. 10, — 57
- iii. 18, — 59, 68
- Amos ix. 13, — 59
- Micah ii. 11, — 60
- vi. 15, — 57
- Nah. i. 14, — 59
- Hab. ii. 5, — 63
- ii. 15, — 64
- Hag. i. 11, — 57
- Zech. ix. 17, — 57
- Matt. ix. 17, — 75
- xi. 18, 19, — 77
- xv. 34, — 86
- xxi. 33, — 79
- xxiv. 38, — 79
- xxvi. 26, — 71, 79
- Mark ii. 22, — 75, 83
- vi. 38, — 86
- xii. 1, — 83
- xiv. 22–25, — 71, 83
- xv. 23, — 83
- Luke i. 15, — 82
- v. 37, — 75, 84
- vii. 33, — 84
- x. 34, — 84
- xii. 19, 45, — 84
- xvii. 27, 28, — 84
- xx. 9, — 84
- xxi. 34, — 84
- John ii. 1–11, — 85, 89
- Acts ii. 13, — 89, 90
- xxiv. 25, — 91
- Rom. xiii. 13, — 92, 103
- xiv. 13, — 92
- xiv. 14–21. — 95
- 1 Cor. vi. 10, 12, — 100
- viii. 9, — 98
- viii. 4–13, — 100
- ix. 25, — 100
- x. 16, — 71
- x. 22–30, — 96, 100
- x. 33, — 96
- xi. 20–34, — 100
- Gal. v. 19–24, — 103
- Eph. v. 18, — 103, 104
- Phil. iv. 5, — 105
- Col. ii. 36, — 105
- 1 Thess. v. 7, — 106
- 1 Tim. iii. 2, 3, 10, — 106, 107
- iii. 8, — 111
- iii. 11, — 112
- iv. 4, — 112
- v. 23, — 108, 116
- Titus i. 6, — 104
- ii. 2, 3, — 117
- 1 Pet. i. 13, — 118
- iv. 1, 5, 7, 104, 118, — 119
- v. 7, 8, — 119
- 2 Pet. i. 6, — 119
- Rev. vi. 6, — 119
- xiv. 8, 10, 19, — 120
- xvi. 19, — 120
- xvii. 2, — 120
- xviii. 3, — 120
- xviii. 13, — 119
- xix. 15, — 120

Seixas, Prof., 12.
Seor, 80.
Sermon, first, 10.
Shakar, 55, 56, 60.
Shanks, G. H., 58.
Shaw, Dr., 19.
Shemarim, 60.
Sick, wine for, 34, 109.
Smart, Prof., 42, 43.
Smith's Antiquities, 29, 34, 37, 45, 46, 76, 90, 110.
Smith, Dr. Eli, 30, 32.
" Wm., Dict., 15, 86, 91.
Social usages, 9.
Sober, 107, 112, 118, 119.
Sorek, 20, 22.
Soveh, 59.
Spain, 23, 41, 44, 49.
Stuart, Prof., 13, 43, 51, 54, 55, 56, 57, 72, 80, 84, 122.
Stum, 40.
Stumbling, 92.
Subsidence, 24, 36.
Succus, 34.
Sulphur, 24, 39, 40, 41.
Supper, Lord's, 79.
Sweet fruits, 18.
" natural taste, 22.
Swift, Judge, 44.
Swineburn, 23.
Syria, wine of, 28, 32, 49.
Syræneum, 59.
Syrup, 16, 25.

Taste, sweet, natural, 22.
Telemachus, 47.
Temple, S. G., 19.
Temperance, 91, 103, 119.
Temperate, 100, 103.
Temperature, Palestine, 19, 21, 22.
Testimony, 12, 121, 122, 123.
Thermopolium, 49, 50.
Thayer, 40, 80.
Theophrastus, 49.
Texts, classified, 12, 62.
Thracian wine, 49.
Tirosh, 57, 58, 68.
Translations of Bible, 53.
Trapp, Dr., 28.
Treat, Capt., 35, 49.
Trench, Dr., 87.
Turner, 17, 81, 114.

Unfermented wine, 40, 49, 51, 56, 79, 90.
Unintoxicating, 35, 42, 109.
Usages, social, 9.
Ure, Dr., 14, 24, 33, 36, 39.

Varro, 48.
Venetians, 30.
Vigilant, 107.
Vim, vi, vires, 33.
Vinous, 16, 17, 18, 25.
Vinum, 51.
Volney, 28.
Virgil, 28, 44.

Walker's Dict., 15.

Warm climate, 18.
Warren, Capt., 22.
Water, 48.
Webster, Noah, Dict., 14.
Wedding, Cana, 85.
Welbeck, 21.
Westein, 117.
Whiston, Wm., 23.
Whitby, Dr., 100.
Wilson, Capt., 22.
Wine, 14, 26, 42, 48, 61.
" bad, 63.
" good, 66.
" new, 75.
" mixed, 50.

Wines, preserved, 25.
Wives, 112.
Women, Roman, 44, 45.
Worcester, 15.
Words, generic 54.
" Hebrew, 58.
" Greek, 60.
" Latin, 60.
" English, 60.

Yahn, Dr., 20.
Yayin, 54, 55, 58, 61, 64.
Yeast, 10, 24.

Zouk, 29.

www.ingramcontent.com/pod-product-compliance
Lightning Source LLC
Chambersburg PA
CBHW022137160426
43197CB00009B/1318